FARM MANAGEMENT
Theory and Practice

By the Same Author

—Emerging Africa: Potential and Challenges
—Natural Resources Management and Agrarian Development

About the Author

Dr. Reji D. Nair is Assistant Professor in the Department of Economics, Arba Minch University, Ethiopia. He was an officer in the Kerala State Planning Board, India. Dr. Nair has published many books including *Africa in the New Millennium: Challenges and Interventions, Land Economics : Issues in Land Use, Natural Resources Management and Agrarian Development* and *Environment Conservation in the Twenty First Century* (Malayalam). He is a recipient of the Dr. B.R. Ambedkar Fellowship Award (2006) of the Bharatiya Dalit Sahitya Academy. He is a notable writer in Malayalam and has published many literary works including children's literature, Science Fiction and Short Stories. Dr. Reji has shown his calibre in the fields of teaching, research, planning and social activities. His other major publications include *Emerging Africa: Potential and Challenges* and *Farm Management : Theory and Practice.*

Farm Management
Theory and Practice

Reji D. Nair

CONCEPT PUBLISHING COMPANY PVT. LTD.,
NEW DELHI-110 059

ISBN-13: 978-81-8069-710-4 (HB)
ISBN-13: 978-81-8069-711-1 (PB)

First Published 2010

Published and Printed by

Concept Since 1974

Concept Publishing Company Pvt. Ltd.
Regd. Office:
A/15-16, Commercial Block, Mohan Garden
New Delhi-110059 (India)
Phones : 25351460, 25351794, *Fax* : 091-11-25357109
Email : publishing@conceptpub.com
Website : www.conceptpub.com

Editorial Office:
H-13, Bali Nagar, New Delhi-110 015, India.

Cataloging in Publication Data--*Courtesy:* D.K. Agencies (P) Ltd. <docinfo@dkagencies.com>

Nair, Reji D.
Farm management : theory and practice / Reji D. Nair.
p. cm.
Includes bibliographical references (p.).
ISBN 9788180697104

1. Farm management. 2. Farm management--Mathematical models. 3. Agricultural productivity. 4. Agricultural productivity--Mathematical models. I. Title.

DDC 630.68 22

Preface

Farming is rapidly changing and even mechanization and computerization are also influencing the way farms are managed. As new technologies develop and world markets emerge; the new global community impacts increasingly upon the farmer. It is still feasible to operate a successful and profitable farm in technologically advanced countries; but only if the right choices are made, proper planning is practiced and good management is adhered to. This book is designed as an up-to-date, broad and balanced perspective on farm management. If the reader is new to agriculture or farming, it will develop an appreciation of the scope and nature of managing a farm, whether it is small or large. For established agriculturists it is a useful way to refresh and update trends and techniques. The book consists of nine chapters.

Farm management is the process by which resources and situations are manipulated over time by the manager of the farm system in trying, with less than full information, to achieve his or her goals. The first chapter is an introduction to Farm Management. It covers areas such as types of farms, definition, features and history of farm management, features of agriculture farm management perspective, and rationale for government intervention in farming.

Major theories underlying farm management are discussed in chapter two. It includes farm-system theory and theory of farm management from objectives. It also covers practice of farm management and brief measures to manage small farms more effectively. Developing country's agriculture characterizes small-scale farmers. There is no magic that can be waved to make the general level of management on small farms more effective overnight. Such improvement should be realistically viewed as an ongoing long-term task involving incremental gains over time.

Farm operations need many inputs including material inputs and services. Farm service sector has some unique features which we cannot see in other economic sectors. Farm service sector includes both purchased and non-purchased inputs. Some of the unique features of the farm service sector such as co-operatives, insurance, farm credit, and farm labour are discussed in the third chapter along with the purchased inputs.

One of the major elements of farm economics is agriculture production economics. Agriculture production is like any other sectors abide by the rules of basic economic theories. Production theory is based on production functions and laws of production, which are reviewed in Chapter four in detail. Production functions, isoquants, technological progress, laws of production including law of returns to scale and law of variable proportions are all covered in this chapter.

One distinguishing feature of agricultural production is that it always involves biological processes. This has many important implications for the application of economic theory to the study of agricultural production, marketing, and pricing. Agricultural production, crop production particularly, requires environmental inputs such as sunlight, water, heat, and time that are often beyond the control of farmers. The control of production quality and yields is often far from complete. In addition, crop production requires land, so the production activities are spread thinly over millions of acres. The application of economic theories in farming is proved in detail in chapter five. It includes application of production function in agriculture, Cobb-Douglas production function, irrational production in agriculture, application of input-output analysis, application of cost functions, family farm model, farm output and input pricing, analysis of farm programmes by giving emphasis to direct payment programme, product choices and production possibilities, farm organizations and budgets, land tenure arrangements, and supply of individual farm products.

Chapter six is an analysis of farm risk management. Risk is an important aspect of the farming business. The uncertainties of weather, yields, prices, government policies, global markets, and other factors can cause wide swings in farm income. Risk management involves choosing among alternatives that reduce the financial effects of such uncertainties. The analysis in this chapter includes types of farm risks, risk management strategies, required risk management programmes, and measurement of farm risks.

Farm business analysis will use various ratios and indicators taken from the financial statements. The most useful function of these tools is to illustrate trends over time. A series of these figures over time shows where the business is heading and help farmers make decisions to alter course. Chapter seven analyses the farm business analysis tools in detail. It includes liquidity measurement, solvency, profitability, farm efficiency and a detailed discussion on cost of production budgeting. It also has an appendix which illustrates the farm business analysis with numerical example.

Planning means deliberate attempt on the part of a country or government or a business organization to follow a specific pattern of growth and structural change. Chapter eight discusses application of important planning techniques such as input-output analysis and linear programming in agriculture. It also reviews the agriculture project planning tools such as net present value, internal rate of return and benefit cost ratio.

As we discussed earlier farming is one of the sectors which is mostly open to frequent risks. It ranges from climatic fluctuation to price variations and the structural transformation of economies. Chapter nine discusses supply response of agriculture from the perspective of price. It covers the Cobb-Web model with illustrations and a simple supply response model. Role of expected price in supply response will be analyzed in detail.

This book will be helpful to undergraduate students of agricultural economics, farm management, farm accounting and rural development. It is also of immense help to farm management practitioners and even literate farmers.

I got a lot of advice and assistance from many of my friends and colleagues while writing this book. I wish to express my deep sense of gratitude to Dr. Radharemanan Pillai, Professor, Department of English, Arba Minch University, Ethiopia and Dr. M.S. Vasu, Assistant Professor, Department of Accounting and Finance, Arba Minch University, Ethiopia for their Solid advices. I owe a deep sense of gratitude to all my colleagues in the Economics Department, Arba Minch University, especially Mr. Asrath Haile, Former Vice Dean, Faculty of Business and Economics and Mr. Tadesse Emiru, Head, Department of Economics for their constant encouragement and co-operation. No less is the gratitude which I owe to all my family members, without their support I could not have completed the present work.

I dedicate this book to my father and mother for their inspiration and support.

Reji D. Nair

Contents

List of Tables

List of Figures

List of Boxes

1

Introduction to Farm Management

Introduction

Management is nothing new. It is something we do every time we choose between alternatives as we attempt to make the most of our lives. Women and men have been necessarily acting as managers ever since the origin of *homo sapiens*. Farming is rapidly changing. Mechanisation and computerisation are just two things which are having a great impact on the way farms are managed. As new technologies develop and world economies merge; the new global community impact increasing upon the farmer. It is still very feasible to operate a successful and profitable farm in technologically advanced countries; but only if the right choices are made, proper planning is practiced and good management is adhered to. The present chapter is designed as an up-to-date, broad and balanced perspective on farm management. If you are new to agriculture or farming, it will develop an appreciation of the scope and nature of managing a farm, small or large. For intermediate specialists it is a useful way to refresh and update trends and techniques.

This is the era of Farm Management (FM), also known as Farm Management Research (FMR). Although the demise of FMR by the late 1970s seems to have gone largely unnoticed by conventional agricultural research, the inherent importance of good economic performance of farms suggests that FSR is effective in supporting farm management practice might well include economics. If this came about, it would constitute the completion of a loop, since Farming System Research was invented largely by agricultural economists. One might ask why the economists left and is there any reason to bring them back. Figure 1.1 shows the relationship between production and management as part of farm system and the economic importance.

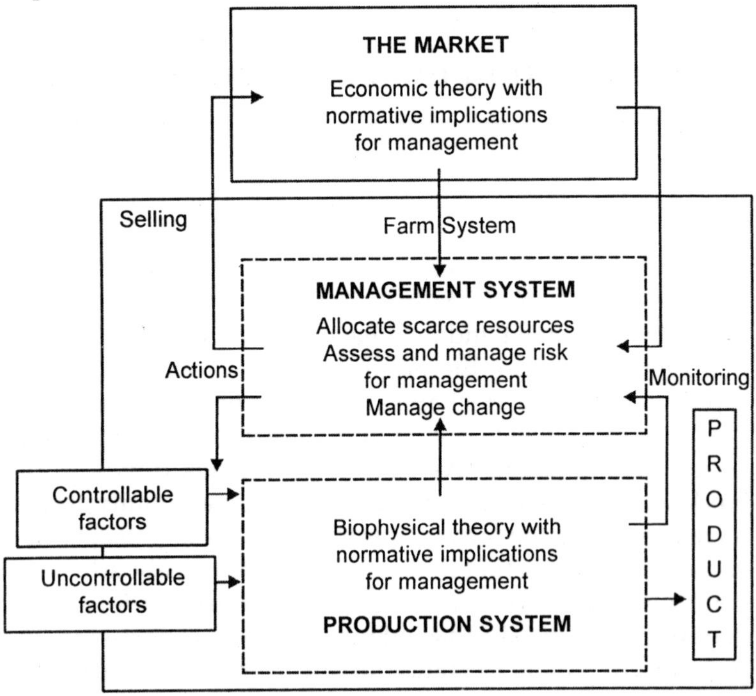

Fig. 1.1 : Farms as System of Production and Management

The distinction between a production system and a management system enables us to distinguish between the technical and the human, or social, element of the farm. Economics is a social science and deals with the management system as it relates to the market. A farm, so represented, suggests that in addition to the role for biological and technical science in FSR in the Production System, there is a role for economics in research aimed at supporting farm performance.

Management involves making decisions, and then following through to implement those decisions. This generally requires long term planning. Planning should be routinely reviewed and modified to fit changing situations. Making good management decisions requires two things:

(i) Full knowledge of the possible options (i.e. What are all of the choices). Lateral thinking helps greatly in being able to devise different and innovative options.

(ii) The ability to foresee all of the important implications arising from following each alternative choice (both short and long term implications). Part of this process involves determining and considering the likely problems associated with each option (i.e. this is called Risk Analysis).

Types of Farms

Farms can be of many types, they might be a fat lamb and wool producing farm, a dairy farm producing milk, or a cropping venture producing vegetables or grain. Most farms, while generally specialising perhaps in one or two areas, are often capable of producing a variety of products without too much changeover. This is a handy position to be in, especially considering the sometimes volatile nature of such factors as the weather and markets. What is produced may not necessarily be for sale, but to reduce costs that would otherwise have to be outlayed, such as hay or fodder crops grown by a dairy farmer to feed his herd in times of poor pasture. Alternatively, having a milking house cow, or running a few head of beef for meat are simple, but effective ways of reducing expenditure.

The first step in farming is the question of what to produce. It is tempting to choose to grow particular livestock or crops because it 'seems' a great idea; without fully considering (and comparing) all other options. Farmers may choose to grow wine grapes because they like drinking wine. Some farmers choose to plant a particular crop, because they know someone else who just had a good year with it. The important steps behind choosing crops or livestock can be listed out as follows :

(i) List all of the possible things which can be grown/ produced in a locality. Be sure to consider factors such as soil types and conditions, water supplies, climatic conditions, etc.;
(ii) Consider what facilities/abilities already have (i.e. land, manpower, expertise, money, etc.). Remove those things from the list of potential products that would require facilities which would be expensive or difficult to provide;
(iii) Consider how long it might take before each potential product would be ready to market. How long would it be before to get a return on the investment? Can we cope financially until then? Consider what demand there might be for each of these things at that point in time. Reduce the list of potential products further;
(iv) Consider the level of risk involved with each of the remaining products. Think about what might go wrong with each (e.g. flood, drought, pest plague, over supply on the market, etc.). Now produce a new list arranging the options in order according to risk—the least risky at the top, the most risky at the bottom; and
(v) Choose what we want to produce. We may decide to produce more than one type of produce—hence reducing the likelihood of serious failure (i.e. if one fails, another may turn a profit).

The use of scientific standards in farming is a relatively new concept in farming in developing countries. The produce of farms might have been tested for impurities or chemical traces but actual farm practice standards were left to the farmer and still largely are. However, more buyers of produce are requiring certain standards in the ways produce is prepared for market. This, they feel ensure the quality of the goods, whether they be eggs, milk, meat or cabbage. Standards might detail the type of housing that animals are kept in, the food they are given, the levels of chemicals they will be subjected to. They will also force farmers to look at their own methods of farming and this will hopefully encourage better farm practices all round.

Due to a number of key factors, such as cost and suitable knowledge, the people who are most likely to become farmers are those people who have been raised or lived on a farm. This is not always the case and recently there have been increasing numbers of people who are leaving urban lifestyles for a farming life. Rich farmers (rarely) often start with a small allotment, while maintaining their mainstream work or income until they are comfortable about making a living from their 'hobby'.

Farming is a dynamic business, don't let the old cliché of the slow talking man on the land fool you. Those farmers who don't keep their finger on the economic pulse soon go out of business. Improving an existing farm requires assessing the positive and negative aspects of what exists on the property, and weighing up our options for improvement. Are the pastures weary? Do the fences need attention? What about the stock, for example, are the cows second rate milkers because no breeding programme has been in place. Obviously the list of questions goes on and on and should continue even when one feels that he has rectified many of the existing problems.

Share Farming

There are many forms of share farming arrangements available and they tend to be individually tailored to the situation. Often they will involve a percentage of the profit in lieu of wages for a farm worker, but may also include other factors such as rental or lodgings, or partial wage plus a smaller percentage of profits, etc. The share farmer may have total managerial control of the property albeit with some budgetary considerations in place. If we are a potential share farmer be sure to thoroughly investigate any proposed arrangements we may be contemplating before sign any binding agreements, in particular ensure that any estimates of future profits are realistic. In addition be sure to put down in the agreement all responsibilities of both parties, including who is responsible for which costs (e.g. rates, services, insurances, cost of materials).

Leasing

Leasing or renting is generally considered a more flexible way of farming than share farming. Farmers can expand or reduce the amount of leased land to match market forces (subject to the availability of land for leasing). It does however have the problem of "insecurity of tenure". Leasing can allow a farmer with limited capital to start a new enterprise. A major problem with leasing land, particularly for short term periods, is that most farmers will be unwilling to spend much time and effort on maintaining or improving the leased property, unless they absolutely have to (e.g. replace fences that are falling down). This can result in a decrease in the productivity of the leased land, often in a very short time. Some lease agreements stipulate certain activities that must be carried out on the part of the person who is leasing the property, such as maintaining fences, controlling weeds, and maintaining pasture quality.

History of Farm Management

It is not surprising, given the necessity for food, to find many references to farm management in the writings that have survived from ancient times. Best known in this regard are the Roman treatises *De Re Rustica* of Cato (234 to 149 B.C.) and *Rerum Rusticam Libritres* of Varro (116 to 28 B.C.). Our favourite, however, is the Greek Hesiod who lived in the Ninth Century B.C. In his *Works and Days* (Evelyn-White, 1914) he gives us what to our knowledge are the first recorded principles of farm management. Few would dispute the good sense of much maxims as "Do not put your work off till tomorrow and the day after, for a sluggish worker does not fill his barn, nor one who puts off his work" or "Avoid shady seats and sleeping until dawn in the harvest season ... Be busy then, and bring home your fruits, getting up early to make your livelihood sure" or "Propitiate [the gods] with libations ... so you may buy another's holding and not another yours".

Since the time of the ancients there has been a continuing stream of writings on management in the context of, in particular, trade, business, warfare, politics and statesmanship. Perhaps best known from the medieval period are two books oriented to management in civil society and politics: Thomas More's (1478 to 1535) *Utopia* and Machiavelli's (1469 to 1527) *The Prince*. Through the Eighteenth and Nineteenth centuries, much of the work of the classical economists such as Adam Smith, James Mill, John Stuart Mill and William Jevons is sprinkled with considerations of the role, function and importance of management. Other notable contributors in this period were Carl von Clausewitz, a Prussian general who wrote on military management; Charles Dupin, a French engineer who emphasized worker welfare and the need for integrity and accountability in management; and Charles Babbage, an Englishman and pioneer of computing, whose work was seminal to the development of the scientific approach to the study of management.

Spurred on by the industrial revolution and the emergence of large-scale organizations (such as railways and multi-branch banks), what is now known as the scientific management movement began in about 1870. Many were involved but, today, Frederick W. Taylor—a U.S. engineer—is credited as being the founder and driving force. His interest was in management in an industrial context, where, in order to maximize output with a given level of effort, he argued that the scientific method had to be applied to job determination, worker selection, creation of a good work environment and specification of output standards so as to determine properly the task for each worker. If the worker then achieved excellence in output, he or she was to be rewarded. If standard output was not achieved, the worker was to be penalized.

Relative to the management of small farms, two points must be made about Taylor's contribution.

(i) It provided the impetus not only for changes in the practice of industrial management but also, since early in the Twentieth century and particularly since Second World War, for the scientific study of management. This area of study has burgeoned in recent decades, particularly in industry but also in the field of farm management. A vast literature and a myriad of texts have been generated, particularly in industrial management. University programmes in management and business administration are now common place and most agricultural faculties, as well as specialist colleges, offer units in farm management.

(ii) Taylor's contribution was in the context of industrial production with its associated large number of employees whose impact on organizational performance was crucial. The same is true of most of the industrial studies of management made since Taylor's initial promulgation of scientific management. For this reason, much of the modem conceptualization of management with its emphasis on the industrial context and a large workforce does not fit well within the context of agricultural production. This is particularly so for small-farm systems using only family labour. Nonetheless, despite the above strictures, there is much in the modem conceptualization of management that is pertinent in a generic sense to farm management.

Definition and Features of Farm Management

To elaborate the general conceptualization of management, consider the following representative sampling of definitions :

(i) "Management aims to accomplish group purposes with the least expenditure of material or human resources" (Koontz, 1969, p. 415).

(ii) "Management aims to ensure that the organization serves its mission in an effective way, and also that it serve the needs of those who control or otherwise have power over the organization (such as its owners, government agencies, unions, pressure groups)" (Mintzberg, 1983, p. 13).

(iii) "Management is the art of getting things done through other people" (Hellriegel and Slocum, 1986, p. 8).

(iv) "Management is the effective and efficient integration and coordination of resources to achieve desired objectives" (Hitt, Middlemist and Mathis, 1989, p. 13).

(v) "Management is the process of working with and through others to achieve organizational objectives in a changing environment. Central to this process is the effective and efficient use of limited resources" (Kreitner, 1989, p. 9).

While not irrelevant to farm management, definitions such as the above clearly do not specifically fit farming situations. From the above definitions we can extract the following features of management :

(i) Successful management is seen as entailing both *efficiency* in resource use and *effectiveness* in the achievement of goals;

(ii) Management is seen as a *universal process* involving a set of *management functions.* These functions are variously elaborated either simply as *planning, organizing and controlling* or in more detail as, e.g., *planning, decision making, organizing, staffing, communicating, motivating, leading, monitoring and controlling.*

(iii) Management is seen as involving a number of *roles.* These are of three types: *interpersonal roles* as figurehead, leader and liaison link; *informational roles* as monitor, disseminator and spokesperson; and *decisional roles* as entrepreneur, disturbance handler, resource allocator and negotiator.

(iv) Relative to its stakeholders, the management of an organization is seen as having three sets of *duties.* First of these is direct supervision and running of the organization. Second is the management of the organization's boundary conditions—its relations with the various environments surrounding it. The third set of duties relates to the development of the organization's strategy which may be viewed as a mediating force between the organization and its environment.

(v) Good management is seen as dependent on the manager possessing *technical, interpersonal, conceptual and diagnostic skills*. Technical skills are those that a manager needs to perform specialized tasks within the organization. Interpersonal skills are those needed to communicate and work well with other people. Conceptual skills relate to a manager's ability to think in the abstract, to see relationships and opportunities, to plan, and to take a global perspective of the organization and its environment. Diagnostic skills are those used to define and understand situations, to recognize problems and see avenues to their solution.

(vi) Successful management is seen as the product of three essential elements, viz., *ability, motivation to manage and opportunity*. Since each of these elements is essential, the relationship between them must be multiplicative.

(vii) Four major approaches to management have been delineated. The *traditional approach* emphasizes order, stability and routine procedures for carrying out the planning, organizing, leading and control functions. This approach perhaps corresponds best to the traditional approach to farm management. The *behavioural approach* emphasizes knowledge about how people behave and why they act as they do. It assumes numerous employees and groups, and is thus less relevant to the management of small farms. The *systems approach* recognizes interrelatedness within the organization and views it as a purposeful (rather than purely mechanistic) system that needs to be considered in a holistic rather than reductionist way. Such an approach is clearly relevant to farm management as has been recognized in recent decades with the development of farm-systems theory and analysis. Lastly, the *contingency approach* recognizes that there is no best way of managing for all situations. Different situations are seen as requiring different management approaches—sometimes traditional, sometimes behavioural and sometimes system-oriented. Such an approach clearly has merit in farm management.

(viii) Management is seen as probably being best described as a mixture of science, practice and art. Successful management, as measured by efficient goal achievement, cannot be guaranteed by the scientific principles of management that are currently available. Insofar as much of management is routine, it is a practice. Not least, successful management is often the outcome of good luck in combination with the artistry of a creative use of knowledge, opportunity and intuition.

(ix) Quantitative methods of analysis can often be highly relevant to some management functions. This is particularly so for planning resource allocation and decision making. However, whether or not quantitative analysis is cost effective will generally depend upon the data required, the complexity of the analysis, the size of the organization or scale of the enterprise and the degree to which its performance is under the manager's control. Simple budgeting may thus be the only cost-effective approach for small enterprises (including small farms) while large enterprises may make full use of all the latest quantitative aids to decision making.

(x) Finally, in this overview of the concept of management, the following question must be asked. Is managerial ability innate and thus dependent on genetics or is it something that can be developed through education and training? Though the relevant genes have not yet been located, there seems to be no doubt that people do differ in their innate managerial ability. At the same time, the plethora of management education programmes available today attests to the view that management ability can be learnt.

Farm Management

Against the above general management background, we can now consider management in the specific context of agricultural production. Consider the following representative sampling of twentieth-century definitions of farm management as proposed in relation to commercial family farms :

(i) "Farm management is concerned with how can the individual farmer so organise the factors of production—land, labour and capital—on his farm, so adapt practice to his particular environment, and so dispose of his product, as to yield him the largest net return, while still maintaining the integrity of his land and equipment" (Butterfield, 1910).

(ii) "Farm management, as the subdivision of economics which considers the allocation of limited resources within the individual farm, is a science of choice and decision making" (Heady and Jensen, 1954).

(iii) "Farm management is concerned with the organization and deployment of the resources put into a farm business—the land, the capital, the labour and that item of overriding importance, the ability and skills of the individual farmer" (Dexter and Barber, 1960).

(iv) "Farm management is concerned with the organization of resources, with planning their use, both within and between enterprises, and with the control of plans both during their implementation and afterwards" (Barnard and Nix, 1973).

(v) "Farm management is concerned with the decisions which affect the profitability of the farm business" (Castle, Becker and Nelson, 1987).

(vi) "Farm management can be thought of as being a decision-making process. It is a continual process... The decisions are concerned with allocating the limited resources of land, labour, and capital among alternative and competing uses. This allocation process forces the manager to identify goals to guide and direct the decision making" (Kay and Edwards, 1994).

There are clearly both similarities and marginal differences among these definitions. It is interesting that the earliest definition from 1910 is little different from the most recent of 1994. The common thread through most of them is decision making about the allocation of resources. Though all the definitions are by respected authorities, we find them inadequate. However, none of them captures all the essential elements of farm management. It is also notable that, reflecting their family-farm orientation, they do not emphasize labour or people any more than other resources. This is in contrast to the general management definitions cited earlier.

If we want to go beyond saying "farm management is what farmers do", is for what—compared with the above definitions—is a more succinct but far more comprehensive definition. This is of farm management as *the process by which resources and situations are manipulated over time by the manager of the farm system in trying, with less than full information, to achieve his or her goals* (Dillon, 1980). This statement highlights the following features of farm management :

(i) Farm management is not farm management research, teaching or consulting;
(ii) The dynamic nature of the farm system and its environment;
(iii) The fact that the farm manager deals not just with resources but also faces the challenge of situations to be met and opportunities to be seized;
(iv) The active role of manipulation as distinct from the more passive role of merely organizing and controlling;

(v) The uncertainty and consequent risk present in the farm manager's decisions, thereby implying attempted (rather than sure) achievement of objectives based on personal preference and subjective judgement;

(vi) By referring to goals rather than profit, the reality of non-profit goals is recognized.

Not only does this definition better capture the challenge and excitement of farm management but, with the deletion of the word 'farm', it serves as an excellent definition of management in general.

Features of Agriculture from Management Perspective

From a management perspective, agriculture is quite distinctive. This distinctiveness primarily relates to the time-dependent biological nature of agricultural production and to the risks this necessarily entails. As well, there are other differences related to cost structure and, in the context of small farms, to the markets faced, their lack of political influence, and the virtually inseparable relationship between the farm system and the farm household. The following are the unique features of agriculture from the management perspective.

Biological Nature

Industrial production, being independent of the natural environment, is mechanical. In contrast, the biological nature of agricultural production causes it to be strongly influenced by the natural environment. In consequence, agriculture has its own innate rhythms and significant elements of agricultural production, which are not under the farmer's control. Species suitability and growth potential, climate, disease and pest occurrence, yield, livestock freewill and gestation period are a few examples. Over time, repetition of the same production process carried out in the same fashion on the same land can give dramatically different results. Nor can production be easily speeded up (or slowed down) in response to favourable (or unfavourable) prices. Indeed, because plant and animal biology are not thoroughly understood, they are still the subject of much research. This can benefit farmers but add to the burden of the farm manager's role if he or she is to keep abreast of relevant research results.

Time Nature

Unlike most other production enterprises, because of its biological nature, agriculture is inextricably time dependent. Compared with industrial production, agricultural production both takes more time and is generally time-related in terms of the seasons. As well, the form of agricultural production, i.e., the species to be grown, is generally dictated by climatic conditions in terms of precipitation, solar radiation etc. Thus, e.g., while grain production requires some three to six months of time and the correct climate and seasons, factory production of a car requires less than a day and is independent of such seasonal elements as rainfall, temperature, day length etc.

From the management perspective, there are the following four influences of time on agricultural production that make it distinctive :

(i) *Flexibility Effect:* By its nature, time gives the farm manager flexibility in the form of options that would not be available if production were instantaneous. Time gives the farmer the opportunity to sequence input injections and/or output harvests in varying ways. Thus the farmer has to decide not only the amount of fertilizer, irrigation water, livestock feed etc. to be used, but also the pattern of its use over the production period. Too, within the time span of any particular activity, decisions about its operation (including marketing of the product, if relevant) can be revised on a sequential contingency basis relative to market or climatic conditions. Obviously, such time-induced flexibility has both advantages and disadvantages for the farm manager. On the one hand it provides the opportunity to correct or improve decisions over time but, on the other hand, it greatly increases the possible volume and complexity of decisions.

(ii) *Time Dependence of Output:* In contrast to industrial production, in many agricultural production activities, time may be a variable input under the manager's control, i.e., a decision variable to be manipulated by the farmer on the basis of its opportunity cost. This is particularly the case for livestock-fattening and aquaculture activities, as well as for some annual crops, and also relative to the replacement decision for perennial crops and woodlots. Often, for most grain crops, it can be ignored because the input of time required, once a particular cultivar has been chosen, is fixed and/or not under the farmer's control.

(iii) *Carryover Effects:* These may be positive as, e.g., when there is a carryover of nutrients from fertilizer application or green manure crops in prior periods. Conversely, they may be negative as in the case of carryover of the eggs of pests or the spores of disease. The extent of such carryover is a function of the time elapsed (e.g., a year or two for green manure but up to 30 years in the case of anthrax spores) and may influence the farmer's current decisions on product choice and input use.

(iv) *Uncertainty:* Time in conjunction with the biological nature and climatic dependence of agricultural production makes farm management a far more risky endeavour than industrial management. In the case of small farms, this is compounded both by the close association between the farm and the household and, if they are market oriented, by their situation as price-takers in both input and output markets. By its nature, time introduces *uncertainty* into farmers' decision making. Because of the complex processes of change, evolution and growth that accompany the flux of time in agriculture and its market and other environments, farmers cannot be sure of the future either in terms of (i) the outcomes of today's decisions or (ii) the opportunities (including new products and technologies) that may become available or (iii) the developments that may occur in their personal, family, social, political, legal, natural or economic environments. Given the information available to the farm manager, the best he or she can do is to make those decisions which are expected to best achieve the desired goals. There can be no guarantee that these risky decisions will, with hindsight, turn out to have been the best that could have been made. Only perfect foresight can guarantee no *ex post* regrets about what might have been.

Resource Portfolio Effect

There are four resource portfolio effects that distinguish farm from industrial management :

(i) *Fixed Costs* in farming are often proportionately higher than in industry. This is due to the importance of land in agricultural production and, on family farms, to the fixed cost of family labour. Because of the more risky nature of agriculture, it is generally more difficult to cover the fixed costs of production in farming than in industry. In consequence, more than industry, agriculture exhibits boom or bust characteristics.

(ii) Relatively higher investment in fixed resources such as land and improvements also causes greater *asset-fixity effects* in agriculture. This fixed investment reduces the opportunity for farm managers, as entrepreneurs, to move into or out of or within agriculture—as may also be the case for small farmers in situations where tenure is on some traditional basis and there is no land market.

(iii) Far more than in industry, farm management has a *custodial or guardianship responsibility* for the natural resources under its control. This is particularly so for land and water which play a more dominant role in agricultural production than in industrial production.

Increasingly, for the benefit of present and future generations, society is demanding that farmers manage their land and water resources so as to enhance or at least sustain their quality rather than to deplete or pollute them.

(iv) Due to the importance of land as a fixed input in agricultural production, the farm manager, more so than his or her counterpart in industry, confronts *the law of diminishing returns.* In consequence, as product and input prices change (as they do more so in agriculture than in industry), the farm manager has an increased volume of decisions to make about how much of each variable input to use and the consequent expected level of output. Concomitantly, the fixity of the land resource and its high contribution to farm costs undoubtedly make it more difficult for management to achieve economies of scale in agriculture than in industry.

Small Farm Nature

Management differences between agriculture and industry are exacerbated in the case of small commercial family farms, which are briefed as follows :

(i) The managers of small farms suffer from a *lack of market power.* They are generally price-takers in both their input and output markets. Often in their input markets they face administered prices. In output markets they generally face few buyers and have to bear the burden of product perishability and the tyranny of distance from the market.

(ii) The close relationship between the farm system and the family on most small commercial farms implies that the farm is not purely a business. The farm and its household constitute an economic entity whose goals are inevitably going to place a heavy emphasis on profit and long term commercial viability but will also involve non-profit elements. Some degree of conflict between farm and family goals is inevitable. The special difficulties for management of working with family members in a business cannot be overstated.

(iii) Compared with industry, the factors of production on small family farms are often not separately supplied: the manager will generally also be a worker, the farm's transport will also be the family transport, and the family home will be the manager's office. This can lead to difficulties in allocation—e.g., to what extent should the farmer spend his or her time as manager or as labourer? Too, on family farms, labour is much less specialized than in industry. The farm operator has to be able to do everything from managing the farm to repairing its infrastructure, keeping its books and ploughing the fields.

(iv) It is a fact of life that the managers of most small businesses, not just farms, have not had the opportunity of formal education and training to develop their managerial skills and capability. Nor, unlike managers in large scale industry, have they been purposefully selected on the basis of their managerial ability. It is, therefore, to be expected that the general level of management in small agricultural enterprises will not be as good as that in larger industrial enterprises. Unless government ensures otherwise, this problem may be compounded by the fact that small farmers are generally disadvantaged in terms of access to information and credit. They also usually lack influence in the setting of agricultural policies and research agendas.

Government Intervention

Government intervention in agriculture required to improve farm economic conditions has taken many forms :

(i) *Adjusting Production to Market*: One approach to improving farm economic conditions is to reduce the number of resources employed to produce a product plagued by surplus conditions. This is achieved by the government's renting whole farms or by paying farmers not to produce the product by requiring them to set aside part of the land normally used to plant this crop in order to qualify for farm Programme benefits. With less land planted to this crop, supply will decline and market price will rise. If the demand curve is inelastic, farm revenue will increase.

(ii) *Price and Income Support Payments*: Another general approach by government to improve farm economic conditions is to directly support farm prices and government purchases of surplus commodities. If the level of production rises during the year, the government can step in and buy excess supply at the announced price floor. An alternative approach is to support farm incomes through direct transfer payments from the government to farmers.

(iii) *Foreign Trade Enhancements*: A third approach to improving farm economic conditions involves the link between agriculture and foreign markets. There are essentially two general ways in which such enhancements can improve farm economic conditions. First, the government can institute tariffs, or a tax on imports, which make imported agricultural products more expensive to domestic consumers, or institute quotas, which limit the quantity of a particular good that can be imported. Both actions protect producers in the domestic agricultural sector. Second, the government can enhance the attractiveness of agricultural products in foreign markets by subsidizing their purchase, thereby stimulating the export demand.

(iv) *Other Forms of Intervention*: The governments world-wide have used a variety of other approaches to enhance economic conditions in agriculture. These approaches include credit subsidies to farm borrowers who cannot otherwise qualify for credit, and loan guarantees for loans made to farmers and ranchers by commercial banks and other non-governmental lenders.

Rationale for Government Intervention

The rationale typically advanced for governmental intervention in the agricultural sector of an economy is the need to :

(i) Support/protect an infant industry,
(ii) Curb market powers of imperfect competitors when necessary to promote social good,
(iii) Provide for food security,
(iv) Provide for consumer health and safety, and
(v) Provide for environmental quality.

Tariffs, a form of tax on imported goods that raises their price to domestic consumers, and import quotas are typically used to shield an infant industry from foreign competition. Many governments intervene in their food and fibre industry to ensure an adequate food supply from domestic sources. Although it is an important issue in all countries, some nations drain off resources from other potential uses in an effort to meet their food needs internally.

In some countries the Environmental Protection Agencies are charged with the responsibility of monitoring the use of the nation's resources, including land (soil erosion), water (pollution), labour (occupational safety) and air (pollution). Thus the farm sector and the rest of the nation's food and fibre industry are heavily influenced by government intervention.

Summary

This is the era of Farm Management, also known as Farm Management Research. Management is seen as probably being best described as a mixture of science, practice and art. Farm management is concerned with how can the individual farmer so organize the factors of production—land, labour and capital—on his farm, so adapt practice to his particular environment, and so dispose of his product, as to yield him the largest net return, while still maintaining the integrity of his land and equipment. In the foregone discussion in this chapter we discussed about various forms of farms, historical origin, definitions and features of farm management and also a vision of agriculture from management perspective. From a management perspective, agriculture is quite distinctive, which is primarily related to the time-dependent biological nature of agricultural production and to the risks it necessarily entails. The chapter concludes with a note on government intervention to adjusting production to market, price and income support payments, foreign trade enhancements, etc. In general we can conclude that farm management is the process by which resources and situations are manipulated over time by the manager of the farm system in trying, with less than full information, to achieve his goals.

REFERENCES

Bennett, D. and Syme, G.J. 1979. *A Report on the Computer Farm Management Planning Service of the Farm Management Service Laboratory of Western Australia* (Inc.) CSIRO Division of Land Resources: Perth.

Collinson, M. 2000. *A History of Farming Systems Research.* CAB International: Wallingford, UK.

Johnson, G.L. 1981. *International Agricultural Programs—Small Farms in a Changing World: Prospects for the 1980s.* Kansas State University: Manhattan.

Makeham, J.P. and Malcolm, L.R. 1993. *The Farming Game Now.* Cambridge University Press: Cambridge, Melbourne.

Martin Upton. 1973. *Farm Management in Africa,* Oxford University Press.

Michael D. Bonehilise and Veronon R. Edanan. 1984. *Farm Management,* Johwlley Sons.

Shaner, W.W.; Philipp, P.F.; and Schmehl, W.R. *Farming Systems Research and Development—Guidelines For Developing Countries.* Westview Press: Colorado.

Taylor, J. 1994. *Fashioning Farmers: Ideology, Agricultural Knowledge and the Manitoba Farm Movement, 1890-1925.* Canadian Plains Research Centre: Regina, Canada.

2

Theory of Farm Management

Introduction

In the context of small farms, the current theory of farm management is best expressed as the integrated consideration of two complementary theoretical frameworks derived from modern management theory. The first of these frameworks is *farm-system theory* with its conceptualization of the farm as a *purposeful system.* This theory constitutes a way of looking at the farm; it provides a checklist of aspects of the farm that should be the concern of management. The second theoretical framework is the overlay of farm system theory by what might be generally called *the theory of management by objectives.* This involves the managerial functions of planning, organizing and controlling the operation of the farm system over time through the use of economic and other principles and administrative procedures. Management by objectives corresponds to the analytical and decision-making activities necessarily undertaken by the farmer in his or her role as manager.

Management by objectives with the aim of maximizing financial profit subject to relevant constraints has been the traditional theoretical approach to commercial farm management. In contrast, farm-system theory has come to formal prominence only over recent decades. However, the theory of farm management by objectives has always to some degree had implicit within it the recognition of the farm as a purposeful system. Both theories could be used as frameworks for the descriptive or positive analysis of a farm's management and performance. Generally, however, they are couched and used as normative theories, i.e., with the aim of providing guidance as to what the farm manager should do if system goals are to be achieved as well as possible. Discussion here is from such a normative perspective.

Farm-System Theory

A *system* is an organized unitary whole consisting of a set of interrelated elements, components or subsystems, each of which is related directly or indirectly to every other element, component or subsystem in the system. With the advent of systems thinking in 1950s, the traditional approaches to understanding phenomena of 'reductionism' (i.e., reducing phenomena to their basic parts and studying these parts in isolation) and 'mechanism' (i.e., that phenomena can be fully explained by mechanical cause-effect relations) have been seen as inadequate for understanding and manipulating systems. Rather, the *systems approach* is based on 'expansionism' or the view that a system is more than the mere sum of its parts. That is, the system's behaviour is not fully deducible from the behaviour of its parts considered in isolation. The systems approach also recognizes that, as well as mechanistic effects, the operation of purposeful systems with their goal-setting and goal-seeking behaviour will be influenced by teleological (i.e. purpose-driven) effects arising from choice and free-will. System performance must, therefore, be judged not simply in terms of how each part works separately, but also in terms of how the parts fit together and relate to each other, and in terms of how the system relates to its environment and to other systems in that environment.

Applied to purposive man-made systems such as farms, the systems approach highlights two features—subjectivity and purpose—that are eschewed by traditional science in its quest for understanding :

(i) A farm system involves not merely the interactions of physical forces but also free-will choices under uncertainty and contests of will arising from the purposiveness of behaviour of animate elements (i.e., people and livestock) in the system. Farm systems thus exhibit non-determinism and capriciousness of behaviour that, before the event, can be captured only by subjective judgement and not by objective fact.

(ii) The understanding (and management) of a purposive system such as a farm necessitates a *teleological or means-end approach* of (i) recognizing the system's goals and (ii) assessing alternative strategies as to how those goals might best be achieved given whatever may be the initial state of the system and its constraining conditions. Use of the systems approach to understand and better manage purposeful systems such as farms, however, is no easy task. Difficulties lie in (a) the free-will and non-deterministic attributes of the farm system; (b) the complexity of purpose and operation that the farm system may possess; and (c) the uncertain dynamic character of the farm system and its environment.

Farm System

Farm-system theory views a farm as a *unique goal-setting* (i.e., purposeful) *open stochastic dynamic artificial* (i.e., man-made) *system* having a major aim of generating income (in cash or kind) for its stakeholders through agricultural production. Generally, a farm system is complex rather than simple. Its purposefulness is ensured by its human and social involvement which enables the system to vary its goals under a given environment. The uniqueness of any particular farm system is guaranteed by its location, history, resources and human elements. In consequence, while the general principles of farm-system management are appropriate for any particular farm, their application must generally lead to different decisions on each particular farm. The openness of the farm system is obvious from its interaction with its environment. The stochastic nature of the farm system is guaranteed both by the free-choice capacity of its human elements and by the stochastic nature of the environment with which it interacts. Necessarily, a farm system is also dynamic by virtue of its purposefulness, openness and stochasticity which ensure that the system changes over time. Too, any farm system is a mixture of abstract and concrete elements or subsystems. The concrete elements are associated with the physical activities and processes that occur on the farm. The abstract elements relate to the managerial and social aspects of the farm.

Farm-system theory sees the farm as having a variety of boundaries or interfaces, i.e., contact points with other systems or environments. Clearly, the physical farm boundary is relevant to the system's land management and is easily specified. Other relevant boundaries will be specified by farm-system interfaces with input suppliers, credit agencies, the local community, government agencies etc. Intra-system resources available to the farm system may be categorized as either physical or non-physical. By their nature, physical resources (such as land, water, buildings, machinery, recreation facilities etc.) are easily specified. Non-physical resources are more difficult to list. They include the knowledge and skills possessed by management, labour and the farm system's social organization. As well, the set of social relationships and the formal and informal organizational structures in which these exist constitute a non-physical resource, as do such attributes as the farm's equity ratio, credit worthiness and the like.

Subsystems of the Farm System

From the perspective of farm-system theory, any farm system—being a purposeful man-made organization—is seen as involving following five major subsystems. These are :

(i) *Technical Subsystem :* Whereby resources, technology, knowledge and opportunity are used to produce agricultural products. The technical subsystem will itself involve a number of production subsystems such as, the cropping system or the irrigation system. The precise nature of these subsystems on any particular farm will be determined by geographic, social, economic and technical constraints. Historically, their interrelationships as subsystems of the farm system have not been adequately recognized. This led to the Post Second World War call for farm management analysis to be on a 'whole-farm basis' which, though an improvement, is usually still inadequate in failing to appreciate sufficiently the farm's non-technical subsystems.

(ii) *Organizational Structural Subsystem :* It corresponds to the formal structure of authority, communication, job descriptions, responsibilities and task allocation within the farm system. On small farms this structural subsystem will not be elaborate.

(iii) *Informal Structural Subsystem :* It exists in any farm system involving two or more persons. The greater the number of persons involved in the farm's social organization (the farm family, workers and their families, neighbours, school teachers etc.), the more complex the informal structural subsystem will be. It is abstract and consists of individuals' motivations and behaviour, informal relationships, feelings of status, power and influence. It also reflects the mixture and interaction of personalities, attitudes, expectations and aspirations of people in the farm system. Regardless of how the formal organizational structure is defined, psychological and social interaction between individuals and groups in the farm system will occur outside the formal structure. Even on small family farms, the 'grape vine' is pervasive—though less so than in large farm systems.

(iv) *Goals and Values Subsystem* : It relates to those goals and values held by the farm system as a purposeful organization. In general, the goals and values subsystem will not correspond exactly to the overall set of (often conflicting) personal goals and values held by individuals in the farm system. The farm's goals and values subsystem will, however, be influenced by the goals and values of individual members since significant disharmony would be destructive to the farm system. The goals and values subsystem will also be influenced by the general socio-cultural and political environment which determines the role individual farm systems are expected to play in society and the requirements they must meet in order to be allowed to function.

(v) *Managerial Subsystem :* It relates to the entire farm system through the farm manager's activity of setting goals, developing long- and short-term plans, specifying organizational structure, deciding on enterprises, choosing technology, allocating resources, seizing opportunities, establishing control processes, harmonizing relations between all the subsystems and with the various environmental supra-systems relevant to the farm etc.—in short, managing the farm system as a purposeful entity. Like the informal structural subsystem and the goals and values subsystem, the managerial subsystem is abstract. It is not concrete in the same way that we can specify the formal organizational structure and write job specifications, or write a manual of technology and a calendar of operations corresponding to the technical subsystem. Components of the managerial subsystem will generally be the persons involved in management (perhaps only one for a small farm system), their knowledge, experience, judgements and preferences, and the information systems used in exercising management.

A generalized schematic representation of the farm system in terms of the above five subsystems is shown in Figure 2.1. The central role of management is clearly emphasized. On large farms all five subsystems would be significant. On small farms involving few people, the organizational structure and informal social subsystems are unlikely to be very important.

Nonetheless, all five subsystems are relevant regardless of farm size; any appraisal of a farm system by its manager or by outsiders should pay due regard to these five subsystems.

Integrating Processes

The five subsystems—technical, structural, informal, goals and values, and managerial—making up the farm system may be viewed as constituting a differentiation of the overall system.

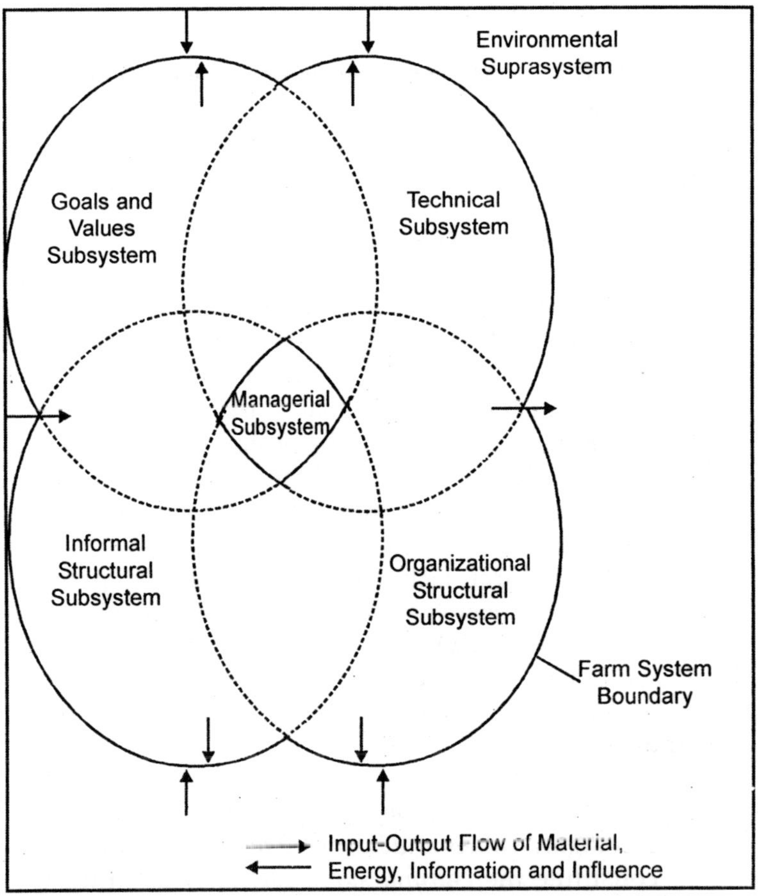

Fig. 2.1 : Generalized Schematic Representation of Farm System

They are the building blocks. The cement which binds these subsystems together or facilitates their interaction and interdependence to form a purposeful whole is a set of integrating variables or processes. These are the *role process, leadership, decision making, information flow and control mechanisms.*

For the farm system to function effectively, its management must pay close attention to these integrating processes. It must provide leadership, carry out effective decision making, ensure that there is a flow of requisite information both for management purposes and to keep others informed, and implement control mechanisms to guide and direct performance. Management must also generally ensure an intra-system environment such that the role performance (i.e., expected behaviour) of persons in the farm system assists rather than hinders system performance.

Farm's Environmental Supra-system

The environmental supra-system is everything external to the individual farm system. The influence of the supra-system environment on the farm system is much greater than *vice versa.* More or less farm system would make virtually no difference to the environment—but it could be profoundly important to those involved with the particular farm system.

The farm system's environmental supra-system involves a number of significant supra-systems of a general nature. These are the cultural, technological, educational, political, legal, demographic, sociological, climatic and economic supra-systems. The state, dynamics, direction of change and interrelations of these environmental subsystems are significant for the individual farm system as a general influence of varying import over time.

(i) *Cultural Supra-system :* The historical background, ideologies, values and norms of the developing society constitute the *cultural supra-system.* This influences views on leadership patterns, authority, rationalism and interpersonal relations within the farm system.

(ii) *Technological Supra-system :* It sets the level of scientific and technological development potentially available to the farm system.

(iii) *Educational Supra-system :* Closely related to technological supra-system and complementary is the *educational supra-system* which determines the general level of literacy and numeracy in the environment, and the availability of specialized services and skills.

(iv) *Political Supra-system :* The general political climate and organization of the enveloping society in terms of political parties, concentration of power and system of government reflects the *political supra-system.* Over time, the individual farm system needs to stay in harmony with the political system.

(v) *Legal Supra-system :* It is specified by constitutional considerations, the nature of the legal system, the jurisdiction of government units and specific laws relating to commercial organization, land use, technology, taxation, labour, health, natural resources, pollution etc. Such laws and regulations, as well as more diffuse legal influences, constrain the operation of every farm system.

(vi) *Demographic Supra-system :* It determines the flow of human resources available to the society in terms of number, regional distribution, age and sex. It also influences the pattern of market demand.

(vii) *Sociological Supra-system :* The class structure, social mobility, definition of social roles and mores, and nature of social organization specify the *sociological supra-system.* These sociological influences from the environment will broadly influence farm system goals and organization, and will determine the social setting within which the farm system operates.

(viii) *Climatic Supra-system :* For nearly all farm systems, the *climatic supra-system* is very important through its close physical interaction with the farm's technical system. At the macro level, global climatic effects can also be significant through their effect on markets and world trade. Significant uncertainty exists about the occurrence, magnitude and impact of global climate change.

(ix) *Economic Supra-system :* Lastly and often most important as a general influence, the *economic supra-system* sets the economic framework both macro (capitalist or socialist, open or regulated etc.) and micro (taxation, commodity policies etc.) enveloping the farm system.

As well as the above general environmental influences, any particular farm system may face other significant environmental influences that are specific to it and must be allowed for by management.

Farm-system Goals

Goals of the farm-system can be seen as arising from three sources. These are :

(i) *Goals of the farm system itself* as a purposive entity (e.g., family sustenance or profit maximization);
(ii) *Social goals* (such as resource conservation) imposed or wished on the farm system by the society in which it functions; and
(iii) *Personal goals* (such as an outdoor life) held by participants in the farm system.

A major task of the farm's management is to harmonize these different goals so as to best achieve farm-system goals. Theoretically, the goals chosen are seen as reflecting need or aspiration circumscribed by *values* or normative views of what is good or desirable. These values will be a composite of those held by farm-system participants and by the members of its developing society.

As a practical matter, it is reasonable to hypothesize or assume that small commercial farms have a number of major or strategic goals. These are likely to include survival, being as profitable as possible, growth of the farm business and conservation of the farm's natural resource base. Obviously, such overarching goals are interdependent. They overlap to some degree and have tradeoff possibilities. Survival, profit and growth are interlinked, with the generation of sufficient profit being a necessary requirement for survival and growth. Conservation of the natural resource base, like the need to meet health, labour, safety and other relevant legal (and socio-cultural) requirements and norms, is best viewed as a constraint. Hence there is logic in the traditional orientation of farm management theory to profit maximization under relevant resource and other constraints. Such an approach also has the advantage of indicating, in terms of forgone profit, the opportunity cost of pursuing non-profit goals. It has the disadvantage, however, of not allowing for the risk inherent in the stochastic nature of the farm system.

To allow for risk considerations, probably the best approach is to view the farm manager as possessing a multi-attribute expected utility function specified in terms of the selected strategic goals, i.e., a function specifying the amount of satisfaction obtainable from the farm system through any combination of goal achievements. Being rational, the manager is assumed to attempt to maximize this utility function, subject to the constraints faced, on the basis of his or her personal judgements and preferences about the risks being faced. This is probably the best technical description of what farm managers actually attempt to do—not in a formal quantitative way but introspectively by means of inductive reasoning from past experience and intuition on the basis of whatever information is available to them. We see this as occurring in much the same way that a football player, without any knowledge of physics or differential equations, knows intuitively how to play the ball. Of course, to the extent that small commercial farms are also family farms, such judgements are further complicated by family goals and interactions between the farm and its household.

Theory of Farm Management by Objectives

Both in theory and in practice the nitty-gritty of management of the farm system is provided through management by objectives. This is done *via* the application of the sequential *management functions* of (as most simply categorized) planning, organizing and controlling across the *management fields* of production, marketing, finance and workforce. Being aimed at the optimal achievement of farm-system goals, the functions of planning, organizing and controlling have a normative orientation. Farm-system goals will be set by the owners or stakeholders of the system. For small farms this will be the farm family or its head, who will often be the farm manager. The actual strategic goals lie at the top of a *means-end schema* in which the achievement of lower-level objectives or sub-goals is an end serving as a means to the attainment of higher-level objectives. Hence, management by objectives relates not just to final objectives or strategic goals but to the whole means-end hierarchy of objectives or goals. Necessarily, therefore, the never-ending cyclical sequential process of planning, organizing and controlling is, in theory, applicable to all levels of the hierarchy of objectives. However, the extent to which this should be so is a decision depending on its benefit-cost ratio at each level. Logic implies that the more important (or the higher) the objective, the more attention management should pay to its planning, organization and control.

The applicability of management by objectives to various levels of the farm system's goal hierarchy necessarily implies its applicability across different time spans, e.g., to short, medium and long-term objectives. The longer the term, however, the greater the uncertainty, the greater the need for management by objectives and the more difficult its successful application.

Planning

The farm system manager has to answer three basic questions: (1) *What* to produce? (2) *How* to produce? (3) *How much* to produce? The answers to these questions constitute the farm's annual *production plan*. This will generally be complemented by other plans, relating, e.g., to farm or personal development. Ideally, the current annual production plan should constitute the first year of a rolling or annually revised multi-year farm plan extending into the future to the point where the net gain from further planning is zero.

Planning decisions about what, how and how much to produce should be based on the basic rule of production economics. Any activity should be undertaken up to and no further than the point where the marginal benefit (i.e., the benefit from undertaking one more unit) of the activity is just balanced by its marginal opportunity cost. In theory, in a world with no uncertainty and with full information about all relevant production relationships, prices and discount rates, optimal annual farm plans could be easily developed (given adequate computer capacity) using the principles of production economics in a multi-year mathematical programming framework specified to accommodate the farm's relevant goals and constraints. These constraints, theoretically, could include not only resource constraints but also technical constraints to ensure that the farm plan meets, if the manager wishes, such criteria as income stability over time, diversity in production, flexibility in product disposal, intra-year dispersion of income, resource sustainability and environmental compatibility.

When uncertainty is present, as is always the case for real-world farm planning, the situation becomes exponentially more complex. Nonetheless, in theory, analytical planning is still feasible so long as information is available about the farm manager's (i) utility function and thus his or her risk preferences, (ii) subjective judgements about the probability distributions of the relevant uncertain outcomes, and (iii) time-preference discount rate. With uncertainty present, choice of plan is a matter of subjective judgement. Rather than choice between varieties of options with sure outcomes, the farm manager faces a set of probability distributions for outcomes corresponding one-to-one to the available set of alternative plans. These probability distributions are specified subjectively by the farm manager, i.e., they express his or her personal degree of belief in the chance of various outcomes occurring. Choice should then be made on the basis of *subjective expected utility theory.*

The presence of time and its corollary, uncertainty, also adds another dimension to farm planning. This is that annual plans may also be revised within years. As well, a contingency approach to planning is possible. This implies that not just one but a number of plans may be specified encompassing what is to be done if various contingencies (such as drought) occur during the planning period. In theory, data used in planning should relate to the period for which plans are being made. Since this is in the future, such data is generally uncertain or taken as unknown. Instead, data from the past, e.g., from farm records of the recent past, are often used. This is seen as providing a link between past control (i.e., monitoring and recording) and future planning. However, naive use of such past data implies the unlikely assumption that the future will be like the past. At both the levels of the farm system and of its technical subsystems, planning and decision making for modern commercial farms have been the subject of much theoretical analysis in recent decades. Stimulated by developments in management science, operations research and computer technology, an array of mathematical programming, simulation and decision theoretic approaches have been proposed and investigated.

Organizing

Compared with planning, organizing is an administrative rather than an analytical or decision-making process. The aim of organizing is to ensure that the farm system's plan is implemented. The plan specifies what, how and how much is expected to be produced. Ensuring that this is carried out requires that necessary tasks are assigned and coordinated, that necessary inputs and arrangements are organized on time, and that appropriate authority is delegated. Two comments can be made about organizing in the context of small commercial farm systems :

(i) The managers of such systems vary greatly in their ability to organize; and

(ii) The capacity to organize is enhanced by having a written farm plan complemented by a technical manual covering the calendar of operations for the products to be produced.

Controlling

In the theory of management by objectives, control is seen as a process of monitoring plan implementation and, as necessary, making adjustments to the plan or its implementation. The information, both experience and data, gathered through control in prior production cycles is seen as contributing to planning for future cycles.

Control may be exercised in more or less elaborate ways. It entails the collection of relevant feedback information, the analysis of such data and, the taking of corrective action. The collection of control data, i.e., the keeping of farm records, has long been advocated by farm management experts and advisers. Many farm recording systems have been promulgated and, these days, software systems are available. These may be more or less detailed but, at minimum, should include major physical and financial information. Analysis of such data is based on its concordance with or deviation from previous performance, expected performance or proposed standard performance. In the 1950s and 1960s, but much less so today, comparative analysis of farm records across groups of similar farms was popular as a guide to management. While helpful to the diagnosis of problems, it is now generally recognized that comparative analysis should be used with caution. Sometimes, deficiencies in plan implementation can be corrected as they are detected. More often, given the farm system's time-dependent nature, adjustments will need to be made to the plan and such corrective action incorporated in future plans.

Practice of Farm Management

The use made either directly or indirectly by small farmers of farm management theory is an empirical question whose answer is sure to change somewhat over time. The question is also likely to be complicated by the fact that small farm systems differ greatly between countries in their historical backgrounds, constraints and environmental supra-systems, and that farm managers differ in their education and training.

Farm-system Approach

The view of the farm as a purposeful open stochastic dynamic system with its various subsystems and supra-systems is a conceptual device. Beyond emphasizing that the farm system should be recognized, understood and managed as a purpose-driven system rather than as just a set of disjointed parts, the systems approach in itself provides no analytical tools aimed at assisting farm management. These come from the application of management by objectives to achieve system goals. Thus, there is no significant gap between farm-system theory and its practice by farmers. Farmers generally have a very keen awareness of their farm as an open stochastic purposeful dynamic system, of its various subsystems and of its environmental supra-systems. Farmers may not have the jargon of systems theory but they do have its insights—and had them long before farm-system theorists.

Farm Management by Objectives

The dichotomy between theory and practice is apparent in farm management texts. Why is there such a large gap between farm management theory and practice? The temptation for theorists is to say farmers are ignorant and irrational. This temptation must be rejected. We know of no evidence to suggest that, within their frame of reference, farmers are irrational. Nor that they are as ignorant as to reject gainful opportunities. A more substantive explanation of the gap is possibly that the suggested techniques of planning and decision making are too difficult and costly for farmers to apply. Too much data is needed and the data is too difficult and time-consuming to collect so that the costs of analysis exceed the benefits. Another possibility is that the analytical products of farm management theory have not been promoted sufficiently well. Given the efforts of government agencies around the world, this appears to be most unlikely. There is also another possibility. This is that the majority of small commercial farmers may judge there to be so much uncertainty in their environment that they cannot exercise sufficient control to warrant the application of sophisticated analytical aids.

Of course, not all small farmers have rejected proposed aids to their managerial decision making. Indeed, with modernization, pressures are such as to enhance the adoption of such aids. World-wide, the business of farming is coming under pressure from the influence of deregulation, open markets and globalization combined with secular decline in agriculture's terms of trade. Significant changes affecting the agricultural system have occurred in most countries in recent years. These changes include : a paradigm shift from production-driven to market-driven agriculture; less government interference; diminishing international trade barriers and increasing international competition combined with rapid transmission of world market volatility; increasing concentration of food processing among transnational and multinational companies requiring consistent quality product; shorter market chains and increased vertical integration; consumer demand for a wide product choice with year-round availability, consistency, convenience and safety; rapid changes in technology, particularly in plant and animal genetics, precision farming, communication and information management; greater integration of chemical, seed and machinery suppliers, and greater emphasis on patents and property rights; increased concern for biodiversity, the environment and personal health; ongoing reduction in the number of farms and a steady decrease in farm size; a greater involvement of women in family farming; and a decline in the 'image' of agriculture.

These changes increase the pressure on small farmers to seek greater efficiencies and be more competitive, as well as external pressure from their credit sources. In seeking greater efficiency small-farm managers cannot adopt sophisticated aids. Rather, we would see them as : (1) becoming knowledgeable about their enterprise gross margins; (2) making greater use of data and information from off-farm sources; and (3) keeping minimal farm records (or having access to minimal standard data) sufficient to develop a basic annual farm plan using gross margins. This broader adoption of the use of gross margins and simple farm planning as being catalysed over time by the use of personal computers and spreadsheet analysis together with the greater involvement of women in the office of the family farm. Only now is it being recognized that farm women constitute a significant potential source of human capital in agricultural development.

Making Small Farm Management more Effective

There is no magic wand that can be waved to make the general level of management on small farms more effective overnight. Such improvement should be realistically viewed as an ongoing long-term task involving incremental gains over time. These gains will accrue as the constraints on managerial performance are lessened. Nor in an unregulated open economic system should it be expected that small farms can escape the pressures of volatility in international commodity markets, ups and downs in income, secular decline in agriculture's terms of trade, and the need to be internationally competitive through ongoing technological improvement and the capturing of size economies.

Solution of the problem of small farm viability lies mainly in small farmers' hands. Those who do not take the right steps to remain competitive, or who suffer bad luck, can expect to lose out under a process of Darwinian selection. Government, however, through its influence on the environment of small-farm systems, can have a significant indirect role in assisting small farmers to be effective and successful managers. This indirect role is over and above any safety-net role that government may play through social welfare policies aimed at assisting non-viable small farmers to be effective in agriculture.

Beyond ensuring that policies do not reflect urban bias, there are following four broad opportunities open to government to help enhance the quality of small-farm management :

(i) Small-farm managers can be helped by the provision of adequate *public infrastructure* of both a physical and institutional nature. Public roads and other facilities such as ports which facilitate market access are probably the most important physical infrastructure. Institutionally, the need is for government to ensure a framework of arrangements that is not detrimental to the provision of communication, credit, electricity, transport, storage, market, research and other services to meet the needs of small-farm systems. These service activities may involve a mix of private and public sponsorship with the latter based, as need be, on the user-pays principle. Research, e.g., might be conducted by both public and private agencies with public-sector research funded in part by output-based levies on producers. The government should also ensure that, except to meet desired social and other objectives, institutional arrangements are not such as to hinder operation of the land market.

(ii) Government needs to ensure that it has *liaison linkages*, preferably of a formal nature, with small farmers through their organizations. Without such liaison linkages, small farmers will continue to be unheard or ignored because of their lack of influence relative to large producers and other sectors of the economy. This is particularly important in the area of publicly funded research. Historically, the evidence is that unless affirmative action is taken to seek out and accommodate the research needs of small farmers, agricultural research agendas are mainly set with an orientation to the wishes of large farmers. This applies just as much to economic and marketing research as to physical crop and livestock research.

(iii) Government should ensure the provision of appropriate and adequate *educational opportunities and facilities* to meet the requirements of small-farm systems. This need for human resource development has a number of dimensions, all of which are of a long-term rather than short-term nature. A primary need is for *basic literacy and numeracy* in the farm-system community. Managers, workers and others associated with small farms in this modern age need to be literate and numerate if they are to make use of modern technology, interpret market signals and remain competitive. *Vocational training* opportunities for existing and potential managers of small-farm systems are also very important. These opportunities may take a variety of forms from short-course arrangements of a few weeks to live-in college programmes offering one, two or three-year courses. Alternatively, they may involve correspondence programmes extending over two or three years with occasional one or two-week residential schools. These courses could be based on application of the management functions of planning, organizing and controlling across the fields of production, marketing, finance and workforce with emphasis on gross margins, budgeting and sensitivity analysis as tools of production planning. Whatever their particular form, such farm management-oriented programmes might be offered by tertiary institutions as an adjunct to their degree programmes or through the Ministry of Agriculture (perhaps as part of extension activities). Lastly, in the educational arena, the government needs to ensure that there is a sufficient output of *agricultural graduates* to service the demands of small-farm agriculture for management information and professional advice whether through research, extension, private consulting or other activities such as policy advice to government.

(iv) The general area of *information* is one in which government can do much to assist the managers of small farms. Traditionally, significant information transfer in agriculture has occurred in many countries through government-sponsored extension services linked to public research agencies. Increasingly, such extension services have moved towards including a management as well as a technical focus. There has also been a realization that relevant and timely information has increasing value as the business of agriculture becomes more complex and volatile. The more situation-specific and time-specific information, the greater its competitive value. Conversely, generalized public information is of decreasing value. Nonetheless, just as can be argued for agricultural research, a case may be argued that a government extension service to small farmers is in the national interest. Such extension should not ignore management information. It could usefully include annual tabulations of enterprise gross-margin budgets on a regional basis, information on expected market needs and outlook on a commodity basis, information on risk-spreading and risk-transfer strategies for handling price risk, meteorological (including probability and forecast) information on rainfall and drought occurrence, and both technical and economic (including risk) information on new technologies and products.

Summary

The current theory of farm management is best expressed as the integrated consideration of two complementary theoretical frameworks derived from modem management theory. The first is *farm-system theory* with its conceptualization of the farm as a *purposeful system* and second theoretical framework is the overlay of farm system theory by what is generally called *the theory of management by objectives.* A farm system is an organized unitary whole consisting of a set of interrelated elements, components or subsystems, each of which is related directly or indirectly to every other element, component or subsystem in the system. The five subsystems—technical, structural, informal, goals and values, and managerial—making up the farm system may be viewed as constituting a differentiation of the overall system and are the building blocks. Both in theory and in practice management of the farm system is provided through management by objectives through the application of the sequential *management functions* of planning, organizing and controlling across the *management fields* of production, marketing, finance and workforce. This chapter also uncovers the dichotomy between farm management theory and practice. Farm systems world-wide are facing drastic changes, and the chapter concludes that it increases the pressure on small farmers to seek greater efficiencies and be more competitive. In seeking greater efficiency small-farm managers should become knowledgeable about their enterprise gross margins; making greater use of data and information from off-farm sources; and keeping farm records sufficient to develop a basic annual farm plan using gross margins.

REFERENCES

Anderson, J.R.; Dillon, J.L.; and Hardaker, J.B. 1985. *Agricultural Systems Research for Developing Countries*. Australian Centre for International Agricultural Research (ACIAR): Canberra.

Bishop, C.C. and Toussaint, W.D. 1958. *Introduction to Agricultural Economic Analysis*, John Wiley & Sons, New York.

Dent, J.B. and Anderson, J.R. 1971. *Systems Analysis in Agricultural Management*. (Ed. J.B. Dent and J.R. Anderson) Wiley: Sydney. pp 3-14.

Dillon, J.L. 1976. *Agricultural Systems*. 1, 5-22.

Doyle, C.J. 1990. *Systems Theory Applied to Agriculture and the Food Chain* (Ed. J.G.W. Jones and P.R. Street). Elsevier Applied Science: London. pp. 89-112.

Kingwell, R.S. and Pannell, D.J. (Eds) 1987. MIDAS, *A Bioeconomic Model of a Dryland Farm System*. Pudoc: Wageningen.

Martin Upton. 1996. *The Economics of Tropical Farming Systems*. Cambridge University Press.

Norman, D. 2000. *A History of Farming Systems Research*. (Ed. M. Collinson) CAB International: Wallingford, UUUK. pp. 293-299.

Ronald D. Kay and William M. Edwards. 1999. *Farm Management*. McGraw-Hill Company.

3

Farm Service Sector

Introduction

Farm operations need many inputs including material inputs and services. Farm service sector has some unique features which we cannot see in other economic sectors. Farm service sector includes both purchased and non-purchased inputs. In this chapter we will discuss about some of the unique features of the farm service sector.

Purchased Inputs

Farmers, like any other producers, buy lots of stuff and by combining it with land, labour, sunshine and rain to produce food. A sample of purchased inputs are given in Table 3.1, which includes cattle feeds, crop seeds, fertilizers, etc., which are mostly purchased from different dealers using various marketing channels. The most important marketing channels include:

(i) Direct Sales

In some instances farmers buy goods and services directly from the producers of those goods and services. For instance, most banking services are purchased directly from the local bank. Direct sales are common in areas such as cattle feed, irrigation systems and special purpose equipments. Direct sales foster and require a close relationship between farmers and the manufacturers of such inputs. The ability to develop and maintain a company owned sales is usually a factor limiting the growth of direct sales firms.

Table 3.1: Purchased Inputs in the Farm Sector

Farm inputs	*Approximate value (%)*
1. Feeds	17
2. Seeds	7
3. Fertilizers	10
4. Pesticides	8
5. Fuel and oil	7
6. Interest	14
7. Labour	24
8. Rent	13

(ii) Agents

The use of franchise agents is a common marketing structure in the farm service sector. In this marketing system the producer of farm inputs signs a marketing agreement with a local merchant to become the local representative of the producer. In this system, what competition exists at the local level is between agents of different manufacturers rather than between different agents of given manufacturer. In this system, the primary marketing efforts of the manufacturer are directed towards the agent rather than to the farmer.

(iii) General Store

Lots of farm inputs such as feed, seed, insecticides and pesticides are sold by farm community general stores. These stores typically buy materials from a variety of vendors to offer their farmers/customers both full service and selection. Such stores are usually locally owned and operated. In many cases, they will more readily extend credit than other input dealers because of their local roots. Frequently, general stores will become franchise agents for a particular item and run the rest of the store parallel to the franchise.

Cooperatives

Processors and farm service suppliers often have local monopoly power in a farm community and use that power to extract monopoly profits from the hapless farmers. One counter strategy has been for farmers themselves to join together in a cooperative association to go to the farm service business. Since these businesses are owned by the farmers, they serve the interest of the farmers.

In the simple terms, a cooperative is an organizational method for doing business. Legally a cooperative is a special type of corporation. There are basically two major differences between Cooperative Corporation and Ordinary Corporation.

(i) A cooperative is designed to do business primarily with its member patrons rather than with the public at large;
(ii) The earnings of a cooperative are distributed to the members on the basis of patronage rather than on the basis of stock ownership.

An agricultural cooperative is a form of business organization set up by a group of individuals in the farm industry for the purpose of performing a service for them. The service they perform cooperatively may be one that was formerly purchased from some other organization, or it may be one that the members of the cooperative formerly did without. The cooperative association is voluntarily owned and controlled by its member-patrons. By combining into a cooperative, farmers gain bargaining power collectively that they would not have individually.

A cooperative exists for the purpose of giving greater economic returns to the members of the organization than could otherwise be achieved. Cooperatives seek to maximize profits, just like any business organization by adjusting output to the level at which the marginal cost of production are equal to the marginal revenue earned. The difference is that profits of a cooperative are distributed to the member-patrons on the basis of patronage rather than to owners on the basis of ownership as is the case in an ordinary corporation. The unique features of cooperatives *vis-à-vis* ordinary business organizations are given in Table 3.2.

Table 3.2: Cooperatives *versus* Ordinary Business Organization

Characteristics	*Ordinary Business Organizations*	*Cooperatives*
1. Ownership of the business	The stockholders	The members who buy stock as they do business with the firm
2. Business management	By management hired by board of directors	By management hired by board of directors
3. Business clients	General public	Member patrons
4. Profit distribution	The stockholders on the basis of share of ownership	The members on basis of patronage
5. Tax payment	Property taxes, corporate income taxes on corporate earnings, stockholders pay personal income tax on dividends distributed from after tax earnings	Property taxes, income tax on earnings not refunded on patronage basis; members pay personal income tax on patronage refunds

Types of Cooperatives

There are three basic types of cooperative associations. The most familiar is the *marketing cooperative,* whose function is to improve the efficiency of the marketing of farm products. Some of the brand name items that may found in any grocery store are processed, packed and marketed by cooperatives.

The Second type of cooperative is the *purchasing cooperatives* designed to take advantage of the bulk lot discounts offered by many manufacturers, and otherwise minimize the cost of procuring materials used in the member patrons' businesses. Most purchasing cooperatives are regional in nature. Most purchasing cooperatives provide services identical with those of a general store, but with higher level of buying power than a local general store can muster.

A third type of cooperative is the *service cooperative,* which provides some service to its members at minimal cost.

Economic Integration

Frequently, a single cooperative association may operate in all three areas. A cooperative that serves as a marketing agency will often mix and sell fertilizer to its members and may also provide a service such as insurance or fertilizer application. This type of economic integration is very common among agricultural cooperatives. Economic integration means tying together of two or more economic activities under some sort of unified control. Economic integration may be a fusion or coordination of business units. There are two main types of economic integration, horizontal and vertical. The term horizontal and vertical are used to convey the characteristics of the type of business coordination.

(i) *Horizontal Integration* : It refers to the general grouping of similar business units under one administrative control. These businesses will be performing the same production or marketing functions at exactly same point in the marketing sequence. Retail food chains represent an example of horizontal integration; where there is a collection of stores brought under one management, each of which is performing exactly the same function. The federation of local grain marketing cooperatives into a regional grain marketing organization provides another example of horizontal integration. The feature of horizontal integration is that a single management gains control over a series of firms performing similar activities at the same level in the production of marketing sequence.

(ii) *Vertical Integration* : It refers to the knitting together of two or more stages of the production and marketing sequence under a single control. It occurs when a firm combines activities that are unlike but sequentially related to those that it currently performs. An example is the food retailer that processes its own dairy or meat products.

When farmers band together in a cooperative association for the purpose of marketing their products, purchasing, or performing some other service for them, we have an example of both horizontal and vertical integration. The act of banding together in an association represents horizontal integration and the farm producers going into the related but unlike activity of marketing, purchasing or service represents vertical integration.

The idea of farmers banding together in cooperative associations for the purpose of achieving market power through horizontal and vertical integration is the basis behind the farm cooperative movement. But because of the extreme economic competition present in the agricultural industry, the market power that could be achieved by the local cooperative associations working independently was negligible. That is, the agricultural industry was so big that even the individual cooperative marketing associations operated essentially as pure competitors. This provides an incentive and an opportunity for still further horizontal integration among local cooperatives. While single cooperative could not afford a massive research programme, for example, an association of cooperatives could make this investment. While a single association did not have enough volume to invest in an installation to process certain by-products, an association of cooperatives could benefit from such an action. As a result, a number of federations of cooperative associations have been formed.

Insurance

Agriculture, as a biological process, is an inherently risky business. The management of this risk is one of the primary responsibilities of the farm managers. Most farmers in developed countries carry property and liability insurance on the farm. But the farmer is exposed to two other risks.

(i) *Production risk* that the crop may fail or the livestock may die;

(ii) *Price risk* that the bottom may fall out of the market leaving the farmer with a negative economic profit.

Insurance is the main solution for managing these risks.

Crop Insurance

Crop insurance quite common in developed countries and is not uncommon in developing countries as well. In crop insurance, the farmer pays a premium to insurance against crop losses. If a loss occurs, the farmer is compensated for the value of the loss. Efforts to provide crop insurance have failed due to many reasons even in advanced countries.

(i) The biggest problem is that any effort to price the premiums at a level that is actuarially sound results in premiums that are so high that only high risk farmers will buy the coverage;
(ii) Another problem is that when there are losses, those losses tend to be catastrophic and widespread, which overwhelms the reserves of the insurance company;
(iii) At the same time when there are catastrophic losses the government will provide disaster relief payments to the farmers, which force the farmers not to take the crop insurance;
(iv) In developing countries majority of the farmers are small-farmers and they don't have the capacity to pay crop insurance premiums.

Income Insurance

Crop insurance covers only production risk; recently in advanced countries pilot programmes have tested insurance that covers both production and price risk. These policies identify an amount of gross receipts a producer is expected to receive from a crop. If the receipts from the crop don't meet the insured amount because of either crop losses or low prices, then the farmer is compensated up to the insured amount.

While income insurance is still in an experimental stage, there is little likelihood that it will be successful in developing countries.

Farm Credit

As we have seen in Table 3.1, around 14 per cent of the total value of the purchased inputs is the share of the interest payments. In developing countries small scale farmers need credit for raising crops as they are subsistent farmers and there is no market surplus. The credit needs of farmers are unique and varied; which can be categorized into long-term, intermediate term and short term.

(i) *Long term credit* : Long term credit needs of farmers are almost exclusively for the purchase of land. The amount to be borrowed is usually quite large and the expected repayment period is quite long. Unlike some other farm credit needs, these loans are secured by the ultimate collateral, the land itself.

(ii) *Medium term credit* : It is used to purchase machinery and equipment, buildings and structures and livestock herds. As with land, the collateral on these loans is usually pretty good.

(iii) *Short term credit* : It is quite common in agriculture. Since the small farmers don't have money in their hand for farming they must borrow to cover planting costs. Such loans are called *production loans*. Most production loans are extended with the understanding that they are to be repaid at harvest time. Since the only collateral for production loans is the crop itself, these loans entail a substantial element of risk on the part of the lender.

Sources of Credit

Farm credit, because of its unique characteristics, is a very specialized part of overall credit system. Most agricultural lenders specialize in credit to agriculture only, and most that don't specialize won't consider making an agricultural loan. Just like any other loan, the lender of money should know the borrower and understand the business of the borrower. As a result several unique credit sources have evolved to serve the needs of the agricultural borrower.

Many farm activities are spread out over time. The adoption of a new technology or a new crop requires investment today, with the payoffs coming on later. Even ongoing productive activity requires inputs in advance, with revenue realized at a later point in time. This is especially true for casual labour or the self-employed; income streams may fluctuate, and such fluctuations will be transmitted to consumption unless they are smoothed through some form of credit. The following two important features of farm credit markets make it problematic.

- It is often very difficult to monitor exactly what is being done with a farm loan. A loan may be taken for an ostensibly productive reason, but may be used for other needs (such as consumption) that cannot be easily transformed into monetary repayment. Alternatively, a loan may be put into a risky productive activity that may fail to pay off. This creates the problem of 'inability to repay' or involuntary default, at which point there is little that a lender can do get his money back.
- The second problem is voluntary or strategic default, a situation in which the borrower can repay the loan, in principle, but simply does not find it in his interest to do so. Such a state of affairs is especially pertinent in contexts where the legal system of loan enforcement is weak.

Agricultural credit markets are usually been served by either institutional or informal lenders.

(i) *Institutional Lenders*: There are the institutional lenders or formal lenders such as government banks, commercial banks, credit bureaus, and so on. Often special banks are set up, as in Thailand, Philippines and India, and in many developing countries, to cater to the needs of agriculture population. The main problem with the formal lenders is that they often do not have personal knowledge regarding the characteristics and activities of their clientele. Often, these agencies cannot precisely monitor just how the loans are used. The problem is not just production *versus* consumption; for example, the fear that a loan taken ostensibly for some productive purposes may be squandered to meet the expenses of a wedding. There are other, more subtle reasons for a systematic divergence between what lenders want and done with the money and what borrowers want.

(ii) *Informal Lenders :* The right sort of informal money-lender may be willing to accept collateral in any form. A large land-owner who has land adjacent to that of a poor farmer may be interested in the tiny plot as collateral (indeed, perhaps more interested in the plot than in getting the loan back). An employer of rural labour will accept labour as collateral, in case the labourer-borrower fails to repay.

It is no surprise, therefore, to find that formal banks cannot effectively reach out to poor borrowers, whereas informal money-lenders (the landlord, the shopkeeper, the trader, etc.) do a much better job. Quite apart from the ability to accept collateral in exotic forms, the informal money-lender often has much better information regarding the activities and characteristics of this clientele.

(iii) *Microfinance:* It is also possible for institutional lending to closely mimic and exploit some of the features of informal lending. For instance, it may be possible to design an innovative rural credit scheme in a rice growing area in which a formal-sector institution acts both as lender and a miller. The combined activity will permit the formal institution to accept rice output as repayment for loans. By mimicking the activities of a trader-moneylender, the formal institution may actually be able to carry out lending activities that reach small borrowers and are profitable at the same time. Even if such mixed institutions are difficult to put into practice, there are ways in which the information base of a community can be put to use by a cleverly constructed lending institution.

Some of the most important features of Microfinance are the following :

- *Group Lending and the Use of Information :* The central feature of Microfinance lending policy is that in the event of a default, no group member is allowed to borrow again. This means that a group has to be formed carefully by the individuals to weed out bad borrowers who could jeopardize the creditworthiness of the group as a whole. The borrowers themselves have the incentive to use this information to form groups, and this induces a form of self-sufficient that no individual-based banking scheme can mimic.
- *Positive Assortative Matching :* Self-selection typically takes the form of positive assortative matching, where good credit risks come together. To be sure, both bad and good risks equally want to team up with the good risks, but if the good risks can identify other good risks, they will form a group together. Thus group formation has the property that it can drive risky types out of the market, because some of the costs of their riskiness are borne by other borrowers instead of the bank alone.
- *Peer Monitoring :* The arguments so far assume that borrowers are intrinsically safe or risky. This is a valid assumption in many situations; group members may not be able to control a fellow member does with her money, although they might still be able to predict what she is going to do (for instance, because they know her personality). In other situations, group members might be able to monitor and influence the choice of

individual projects. In that case, the group will want these projects to be safe, relative to the safety levels they would choose as individual borrowers.

Farm Labour

Labour is one of the farmer's largest purchased inputs. While many farms are self-sufficient with family labour, the trend is towards more and more purchased/hired labour. In developing countries family labour occupies the major contributor of small-holder farm families. Following are the important categories of farm labour, which are used vastly in world agriculture.

(i) *Tenant labour*: Tenants are farm operators who do not own their land. They rent the land from the land owner. Approximately 10-15 per cent of world agriculture is operated through tenant labour. The managerial responsibilities of owner and operator (Landlord and renter) are negotiated between the two parties. There are two common types of rental agreements. In *cash rent agreement* a specific amount of rent is agreed upon between the two parties. In *share agreement* the land owner claims a share of the crop, at the same time the owner share some of the input expenses. In share rent agreement, what the owner will earn in any given year depends on the success of the tenant's farming and market prices of the crop.

(ii) *Labour contracts*: An increasing proportion of the hired labour in advanced countries is provided by labour contractors. These intermediaries are very common in the fruit and vegetable sectors where substantial quantity of seasonal, usually migrant labour is required. The farmer signs a contract with the contractor to provide a certain quantity of labour at a certain time. The contractor then performs a number of tasks that the farmer really does not want to mess with. By using a labour contractor the farmer is able to avoid many hassles and shift the legal liability for labour law and immigration compliance onto the contractor. It is the ability to shift legal liability that makes working with labour contractors so attractive to the farmer needing seasonal labour.

(iii) *Custom hire* : Custom hire refers to the practice of hiring labour and equipment to perform some tasks on the farm. The most common form of custom hire is to hire some one with a combine to harvest the crop. In some cases this may be a neighbour who happens to own a combine who will do custom work for farmers who don't. In other case it may be a small firm with multiple combines that hires drivers and hires out their services. The latter one is quite common in the wheat belts of United States.

(iv) *Consultants*: Farmers use expert consultants in specific tasks of farm activities mainly in developed countries. Farmers' world-wide have used legal and financial consultants for may years, but now there are evolving new types of consultants who respond to the needs of the farmers in an increasingly complex regulatory environment and to the needs of the bottom line. Environment management, pest management and waste management consultants are giving concrete solutions to many of the problems of the farmers. Crop advisers are consultants who work with farmers providing agronomic advice such as seed selection, fertilizer rates and the new field of precision farming. Precision farming refers to adjusting application rates of fertilizers and other chemicals to the specific needs of the spot in the field where the application is being made.

(v) *Family labour* : In small holder farms family labour occupies major contributor of labour power as manager, worker as well as consultants. In developing countries family labour is the backbone of traditional farming. Due to population pressure and land fragmentation small farms have become uneconomical these days and so farmers are not in a position to hire labour. On such situations the farm operations are managed through family labour. However, in recent years socio-economic progress has led to labour migration from farms to cities which complicate the farm labour structure.

Summary

Farmers, like any other producers, buy lots of inputs and by combining it with land, labour, sunshine and rain to produce food. The purchased inputs include cattle feeds, crop seeds, fertilizers, etc., which are mostly purchased from different dealers using various marketing channels. Along with purchased inputs farmers use many farm services also such as cooperatives, insurance, farm credit, etc. An agricultural cooperative is a form of business organization set up by a group of individuals in the farm industry for the purpose of performing a service for them. By combining into a cooperative, farmers gain bargaining power collectively that they would not have individually. This chapter also discussed about the need for insurance services in farming. Crop insurance covers only production risk; recently in advanced countries pilot programmes have tested insurance that covers both production and price risk. In developing countries small scale farmers need credit support as well, since they are subsistent farmers and there is no market surplus. This chapter concludes with a discussion on various forms of labour used in farming.

REFERENCES

Abbott J.P. Makeham. 1990. *Agricultural Economics and Marketing in the Tropics*, Longman Group.

Evan Drummond, H. and John, W. Goodwin. 2001. *Agricultural Economics*, Prentice Hall, New Jersey.

John B. Penson, Oral Capps and C. Parr Rosson. 2002. *Introduction to Agricultural Economics*. Prentice Hall: New Jersey.

Ladd Hollist, W. and F. Lamond Tullis (eds). 1987. *Pursuing Food Security: Strategies and Obstacles in Africa, Asia, Latin America and the Middle East*. Lynne Rienner Publishers. London.

Musgrave, W.F. 1976. *Australian Journal of Agricultural Economics*, 20:3, 133-144.

Pannell, D.J., Malcolm, L.R. and Kingwell, R.S. 2000. *Agricultural Economics*, 23:1, 69-78.

Uma Lele. 1984. *The Design of Rural Development: Lessons from Africa.* World Bank Research Publications. Washington.

4

Theory of Production

Introduction

One of the major elements of farm economics is agriculture production economics. Agriculture production is like any other sectors abide by the rules of basic economic theories. Production theory is based on production functions and laws of production. In this chapter basic economic theories of production are discussed and in the next chapter its application at farm level will be reviewed.

Production Function

The production function is a purely technical relation which connects factor inputs and outputs. It describes the laws of proportion, which is the transformation of factor inputs into products (outputs) at any particular time period. The production function represents the technology of a firm or an industry, or the economy as a whole. The production function includes all the technically efficient methods of production. The production function in traditional economic theory assumes the form $X = f(L, K, v, \gamma)$. The factor's returns to scale, refer to the long run analysis of the laws of production. The efficiency parameter refers to the entrepreneurial organizational aspects of production. Graphically, the production function is usually presented as a curve on two dimensional graphs. Changes of the relevant variables are shown either by movements along the curve that depicts the production function, or by shifts of this curve. The most commonly used diagrams of a single product are shown in Figures 4.1(a) and 4.1(b).

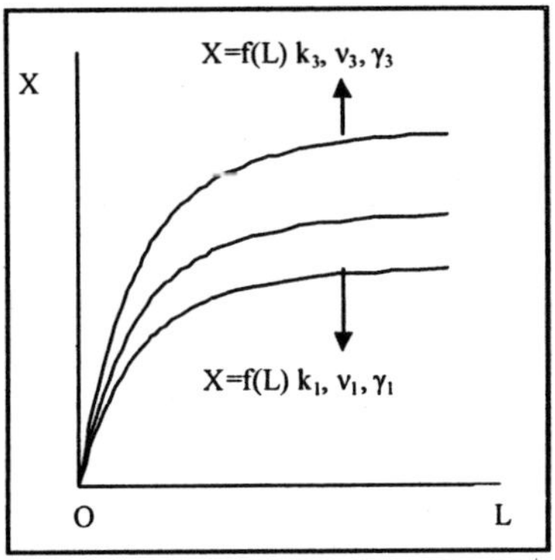

Fig. 4.1(a) : Production Function [X = f(L)]

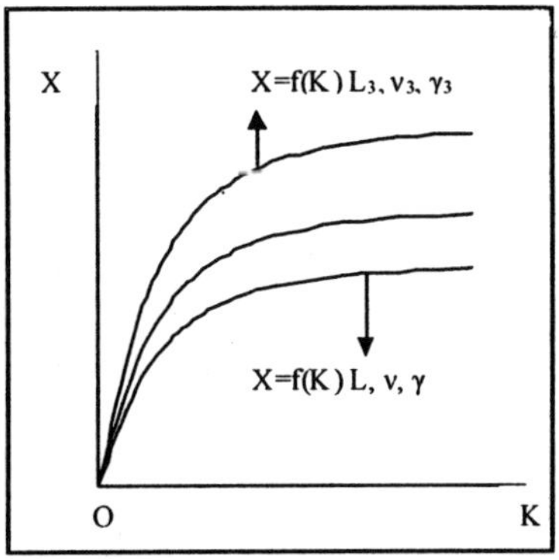

Fig. 4.1(b) : Production Function [X = f(K)]

Each curve shows the relation between X and L for given K, v, and γ. As labour increases, we move along the curve depicting the production function. If capital increase, the production function $X= f(L)$ moves upwards. The *marginal product* of a factor is defined as the change in output resulting from a very small change of this factor, keeping all other factors constant. Mathematically the marginal product of each factor is the partial derivative of production function with respect to this factor.

$$MP_L = \frac{\partial X}{\partial L} \text{ and } MP_K = \frac{\partial X}{\partial K}.$$

Graphically the marginal product of labour is shown by the slope of the production function $X=f_1(L)_{K,\ v,\ \gamma}$ and the marginal product of capital is shown by the slope of the production function $X = f_2\ (K)\ _{L,\ v,\ \gamma}$. The theory of production concentrates on levels of employment of the factors over which their marginal products are positive but decrease.

A method of production is a combination of factor inputs required for the production of one unit of output. Usually a commodity may be produced by various methods of production. For example, a unit of commodity x may be produced by the following process :

	Process A	Process B	Process C
Labour Units	2	3	1
Capital Units	3	2	4

Activities may be presented graphically by the length of lines from the origin to the point determined by the labour and capital inputs. The three processes above are shown in Figure 4.2. The method of production *A* is technically efficient relative to any other method *B*, if *A* uses less of at least one factor and no more from the other factors compared with *B*. Method *B* is technically inefficient as compared with *A*. The basic theory of production concentrates only on efficient methods. Inefficient methods will not be used by rational entrepreneurs.

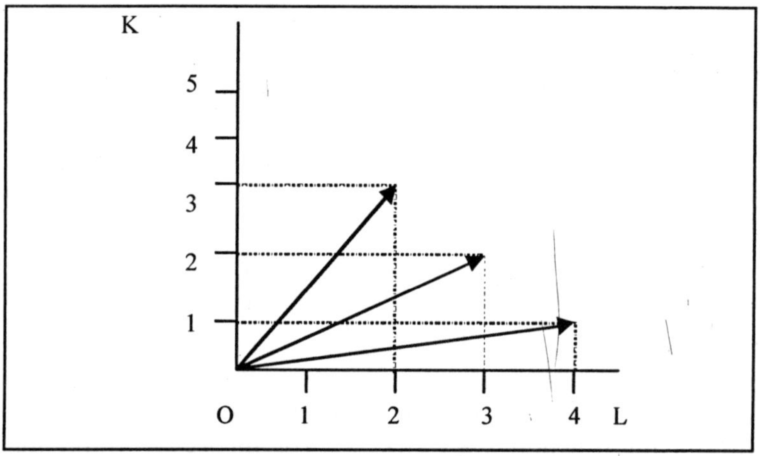

Fig. 4.2 : Methods of Production

Process *A* uses less of some factor(s) and more of some other(s) as compared with any other process *B*, then *A* and *B* cannot be directly compared on the criterion of technical efficiency. For example, the activities are not directly comparable. Both processes are considered as technically efficient and are included in the production function (the technology). When one of them will be chosen at any particular time depends on the prices of factors. The theory of production describes the laws of production. The choice of any particular technique is economic one, based on prices, and not a technical one. A technically efficient method is not necessarily economically efficient. There is difference between technical and economic efficiency.

	A	*B*
Labour units	2	1
Capital Unit	3	4

Isoquants

An *isoquant* includes all the technically efficient methods (or all the combinations of factors of production) for producing a given level of output. The production isoquant may assume various shapes depending on the degree of substitutability of factors. They are broadly classified as following :

(i) *Linear isoquant*. This type assumes perfect substitutability of factors of production : a given commodity may be produced by using only capital, or only labour, or by an infinite combination of *K* and *L*. See Figure. 4.3(a).

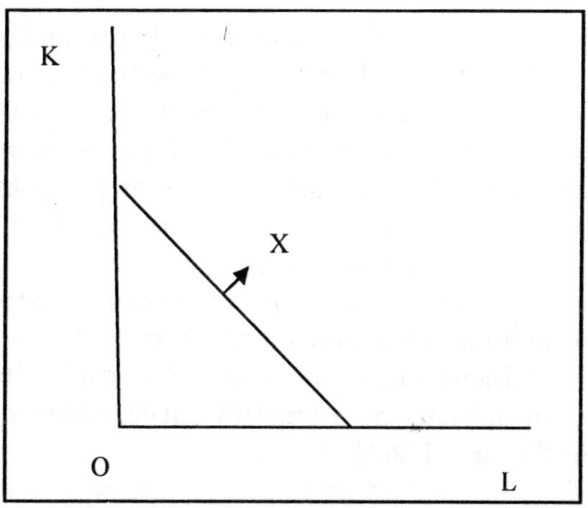

Fig. 4.3(a) : Linear Isoquant

(ii) *Input-output isoquant* assumes strict complementarity (that is zero substitution) of the factors of production. There is only one method of production for any one commodity. The isoquant takes the shape of right angle [Figure. 4.3(b)]. This type of isoquant is also called 'Leontief isoquant' after Leontief, who invented the input-output analysis.

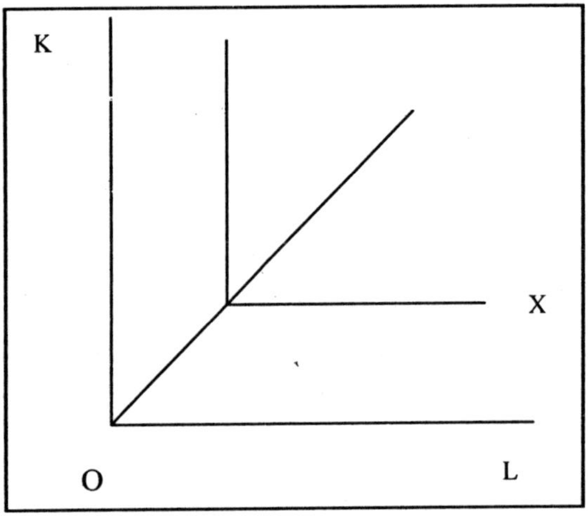

Fig. 4.3(b) : Input-Output Isoquant

(iii) *Kinked isoquant* assumes limited substitutability of *K* and *L*. There are only a few processes for producing any one commodity. Substitutability of the factors is possible only at the kinks. This form is also called 'activity analysis isoquant' or 'linear programming isoquant', because it is basically used in linear programming [Figure. 4.3(d)].

(iv) *Smooth, convex Isoquant* assumes continuous substitutability of *K* and *L* only over a certain range, beyond which factors cannot substitute each other. The isoquant appears as a smooth curve, convex to the origin [Figure. 4.3(c)].

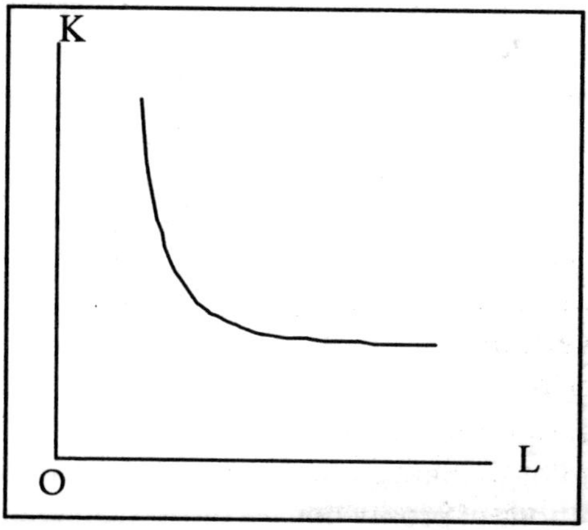

Fig. 4.3(c) : Convex Isoquant

The production function describes not only a single isoquant, but the whole array of isoquants, each of which shows a different level of output. It shows how output varies as the factor inputs changes [Figure 4.3(d)].

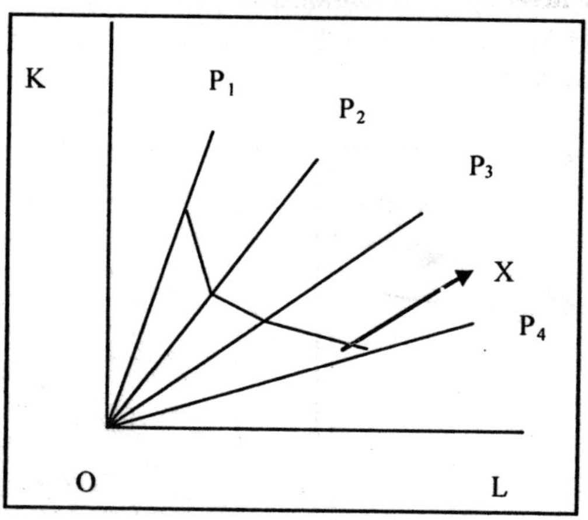

Fig. 4.3(d) : Linear Programming Isoquant

Technological Change

Inputs to production are called factors of production. Factors of production are often classified into broad categories such as land, labour, capital, and raw materials. Capital goods are those inputs to production that are themselves produced goods. Capital goods are machines of one sort or another : tractors, buildings, computers, or whatever. Sometimes capital is used to describe the money used to start up or maintain a business. Financial capital is the term that used to describe this concept of capital goods, or physical capital, for produced factors of production. Inputs and outputs as being measured in *flow* units : a certain amount of labour per week and a certain number of machine hours per week will produce a certain amount of output per week.

Technology changes as knowledge of new and more efficient methods of production becomes available. In addition, new inventions may result in the increase of the efficiency of all methods of production. At the same time some techniques may become inefficient and drop out from the production function. These changes in technology constitute technological progress. Graphically the effect of innovation in processes is shown with an upward shift of the production function [Figure 4.4(a)], or downward movement of production isoquant [Fig. 4.4(b)]. This shift shows that the same output may be produced by less factor inputs, or more output may be obtained with the same inputs.

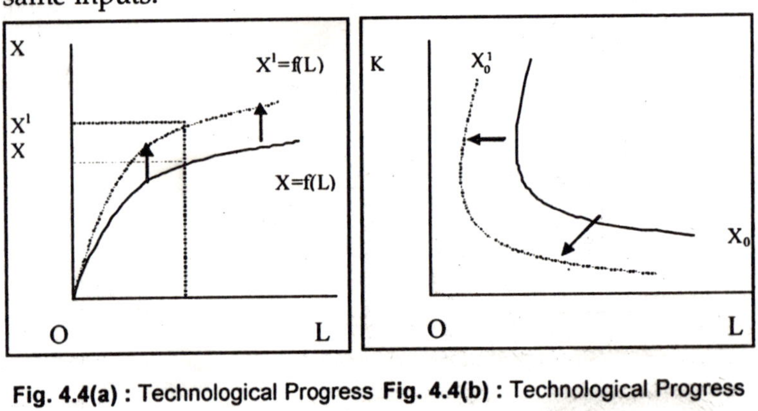

Fig. 4.4(a) : Technological Progress **Fig. 4.4(b)** : Technological Progress

Technical progress may also change the shape of isoquant. Hicks has distinguished three types of technical progress, depending on effect on the rate of substitution of the factors of production.

Capital-deepening Technical Progress

Technical progress is capital-deepening or capital using, along a line on which *K/L* ratio is constant, the $MRS_{L,K}$ increases. This implies that technical progress increases the marginal product of capital by more than the marginal product of the labour. The ratio of marginal products (which is the $MRS_{L,K}$) decreases in absolute value. But taking into account that the slope of the isoquant is negative, this sort of technical progress increases the $MRS_{L,K}$. The slope of the shifting isoquant becomes less steep along any given radius. The capital-deepening technical progress is shown in Figure 4.5.

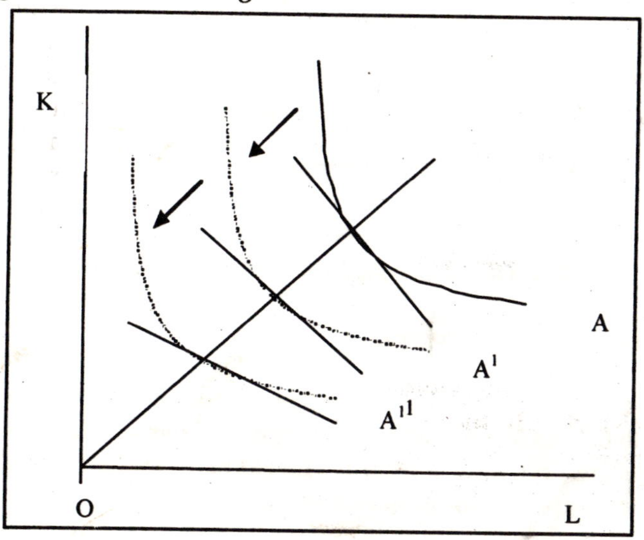

Fig. 4.5 : Capital-deepening Technical Progress

Labour-deepening Technical Progress

Technical progress is labour deepening, if along a radius through the origin with constant *K/L* ratio, the $MRS_{L,K}$ increases. This implies that the technical progress increases the MP_L faster than the MP_K. Accordingly the $MRS_{L,K}$, being the ratio of the marginal products $[\partial X/\partial L]/[\partial X/\partial K]$, increases in absolute value. The downward-shifting isoquant becomes steeper along any given radius through the origin. This is shown in Figure 4.6.

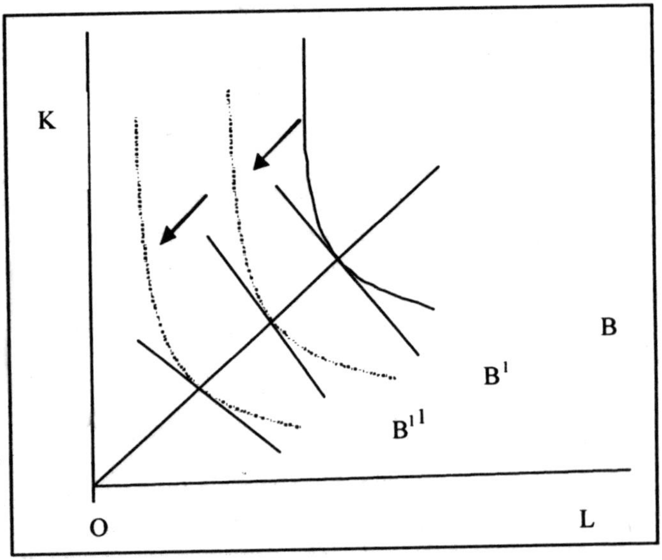

Fig. 4.6 : Labour-deepening Technical Progress

Neutral-technical Progress

Technical progress is neutral if it increases the marginal product of both factors by the same percentage, so that the $MRS_{L,K}$ (along any radius) remains constant. The isoquant shifts downward parallel to itself. This is shown in Figure 4.7.

Properties of Technology

Technologies are monotonic; if we increase the amount of at least one of the inputs, it should be possible to produce at least as much output as we were producing originally. This is sometimes referred to as the property of *free disposal*: if the firm can dispose costlessly any of inputs, having extra inputs around cannot disturb it. Generally *technology is Convex*. This means that if we have two ways to produce y units of output, (x_1, x_2) and (z_1, z_2) when their weighted average will produce at least y units of output. One argument for convex technologies is that if we have a way to produce one unit of output using L_1 units of factor (Labour) one and K_2 units of factor (capital) two and that we have another way to produce one unit of output using L_2 units of factor one and K_2 units of factor two. These two ways are called as production techniques.

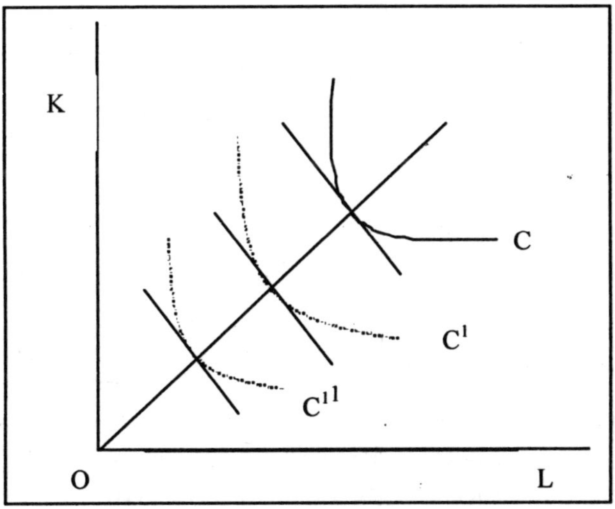

Fig. 4.7 : Neutral Technical Progress

Furthermore, let us suppose that we are free to scale the output up to arbitrary amounts so that (100*a*1, 100*b*2) will produce 100 units of output. But now note that if we have (Technique *A*) $25L_1+75L_2$, units of factor one (Labour) and (Technique *B*) $25K_1+75K_2$ units of factor two (Capital) we can still produce 100 units of output: just produce 25 units of the output using the *"a"* technique and 75 units of the output using *"b"* technique. This is depicted in Figure 4.8 by choosing the level at which we operate each of the two activities, we can produce a given amount of output in a variety of different ways. In particular, every input combination along the line connecting

($100b1$. $100b2$) will be a feasible way to produce 100 units of output.

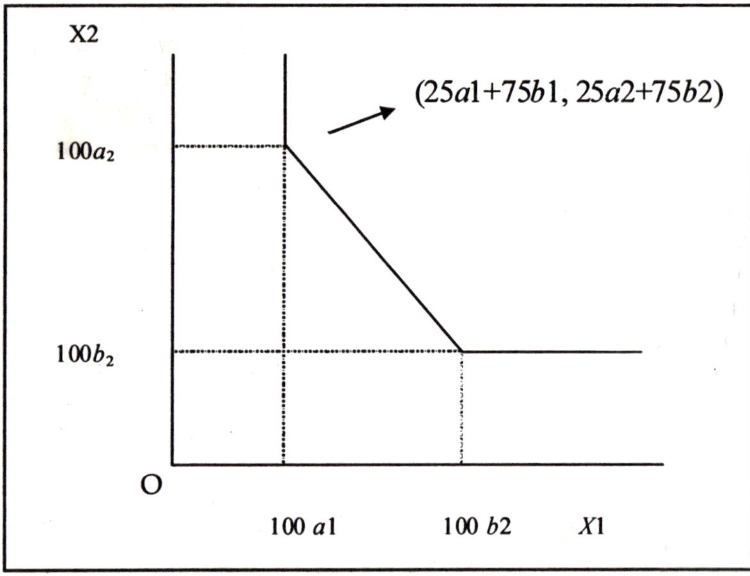

Fig. 4.8 : Convexity of Isoquant

In this kind of technology, where we can scale the production process up and down easily and where separate production processes do not interface with each other, convexity is a very natural measure.

Laws of Production

The laws of production describe the technically possible ways of increasing the level of production. Output may increase in various ways. Output can be increased by changing all factors of production. Clearly this is possible only in the long run. Thus *the laws of returns to scale* refer to the long run analysis of production. In the short run output may be increased by using more of the variable factor(s), while capital (and possibly other factors as well) are kept constant. The marginal product of the variable factor(s) will decline eventually as more and more quantities of this factor are combined with the other constant factors. The expansion of output with one factor (at least) constant is described by the *law of diminishing returns* of the variable factor, which is often referred to as *the law of variable proportions*.

Product Lines

To analyze the expansion of output we need a third dimension, since along the two dimensional diagram we can depict only the isoquant along which the level of output is constant. Instead of introducing a third dimension it is easier to show the change of output by shifts of the isoquant and use the concept of *product lines* to describe the expansion of output. *Product Line* shows the (physical) movement from one isoquant to another as we change both factors or a single factor. A product curve is drawn independent of the prices of factors of production. It does not imply any actual choice of expansion, which is based on the prices of factors and is shown by the expansion path.

The product line describes the technically possible alternative paths of expanding output. What path will actually be chosen by the line will depend on the prices of factors. The product curve passes through the origin if all factors are variable. If only one factor is variable (the other being kept constant) the product line is straight line parallel to the axis of the variable factor (Figure 4.9). The *K/L* ratio diminishes the product line.

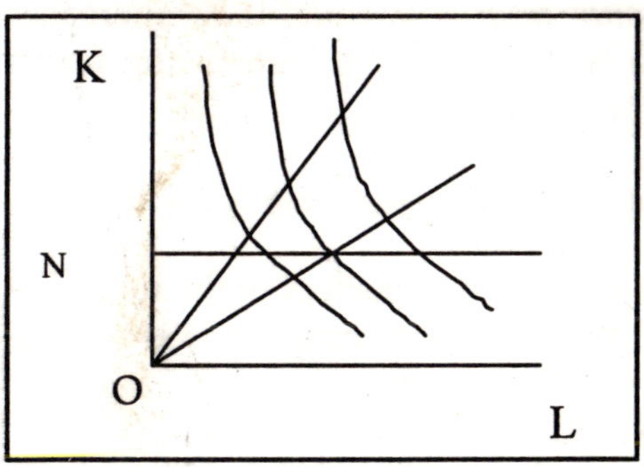

Fig. 4.9 : Product line for *K* given *L*

Among all possible product lines of particular interest are the so called *isoclines. An isocline is the locus of points of different isoquants at which the marginal rate of substitution (MRS) of factors is constant.* If the production function is homogeneous the isoclines are straight lines through the origin. Along any of one isocline the *K/L* ratio is constant (as the *MRS* of the factors).

Laws of Returns to Scale : Long-run Analysis of Production

The long run expansion of output may be achieved by varying all factors. In the long run all factors are variable. The laws of returns to scale refer to the effects of scale relationships. In the long run output may be increased by changing all factors by the same proportion, or by different proportions. Traditional theory of production concentrates on the first case, that is the study of output as all inputs change by the same proportion. The term *returns to scale* refers to the changes in output as all factors change by the same proportion.

Suppose we start from an initial level of inputs and output.

$X_0 = f(L, K)$ and we increase all the factors by the same proportion *k*, we will clearly obtain a new level of output X^*, higher than the original level X_0,

$X^* = f(kL, kK)$

If X^* increases by the same proportion *k* as the inputs, we say that there are *constant returns to scale.* If X^* increases less than proportionally with the increase in the factors we have *decreasing returns to scale.* If X^* increases more than proportionally with the increase in the factors, we have *increasing returns to scale.*

Returns to Scale and Homogeneity of the Production Function

Suppose we increase both factors of the function

$X_0 = f(L,K)$ by the same proportion *k*, and we observe the resulting new level of output *x*.

$X^* = f(kL,kK)$

If *k* can be factored out (that is by taking out of the brackets as a common factor), then the new level of output X^* can be expressed as a function of *k* and the initial level of output

$$X^* = k^v f(L,K) \text{ or } X^* = k^v . X_0$$

and the production function is called homogeneous. If *k* cannot be factored out, the production is non-homogeneous. Thus a homogeneous function is a function such that if each of the inputs is multiplied by *k*, then *k* can be completely factored out of the function. The power *v* of *k* is called the degree of homogeneity of the function and is a measure of the returns to scale. If $v = 1$ we have constant returns to scale. This production function is sometimes called linear homogeneous. If $v < 1$ we have decreasing returns to scale; and if $v > 1$ we have increasing returns to scale.

Graphical Representation of the Returns to Scale for a Homogeneous Production Function

The returns to scale may be shown graphically by the distance (on an isocline) between successive multiple level of output isoquants, that is isoquants that show levels of output which are multiples of some base level of output, e.g., *X, 2X, 3X,...* etc.

(i) *Constant Returns to Scale*: Along any isocline the distance between successive multiple isoquants is constant. Doubling the factor inputs achieve double the level of the initial output; trebling inputs achieves treble output, and so on (Figure 4.10).

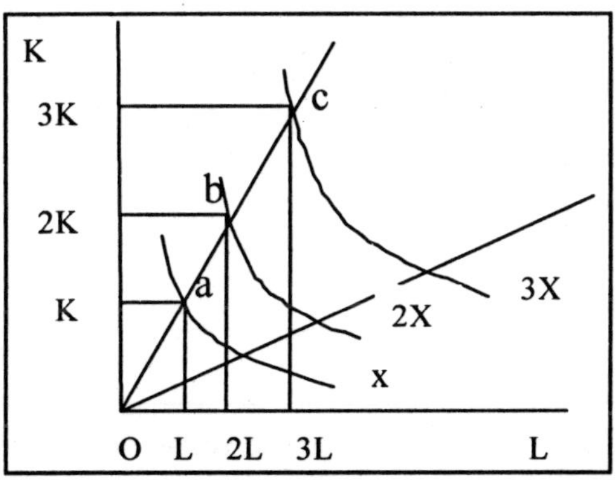

Fig. 4.10 : Constant Returns to Scale: *Oa* = *Ob* = *Oc*

(ii) *Decreasing Returns to Scale:* The distance between consecutive multiple isoquants increases. By doubling the inputs, output increases by less than twice its original level in Figure 4.11 the point *a* defined by 2*K* and 2*L*, lies on an isoquant below the one showing 2*x*.

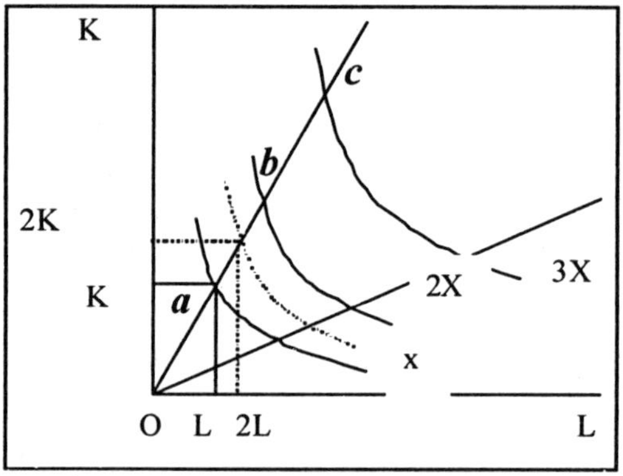

Fig. 4.11 : Decreasing Returns to Scale : *Oa* < *Ob* < *Oc*

(iii) *Increasing returns to scale.* The distance between consecutive multiple-isoquants decreases. By doubling the inputs, output is more than doubled. In Figure 4.12 doubling *K* and *L* leads to point *b* which lies on an isoquant above the one denoting 2*x*.

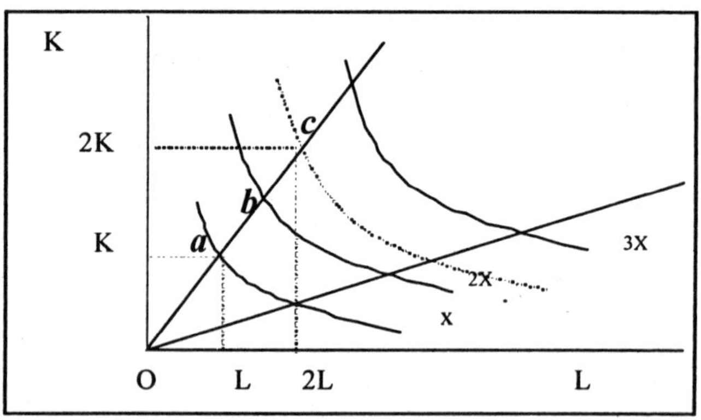

Fig. 4.12 : Increasing Returns to Scale : *Oa* > *Ob*> *Oc*

Returns to scale are usually assumed to be the same everywhere on the production surface, that is the same along all the expansion product line. All processes are assumed to show the same returns over all ranges of output; either constant returns everywhere, decreasing returns everywhere, or increasing returns everywhere. However, the technological conditions of production may be such that returns to scale may vary over different ranges of output. Over some range we may have constant returns to scale may vary over different ranges of output.

With a non-homogeneous production function, returns to scale may be increasing, constant, or decreasing, but their measurement and graphical presentation is not as straight forward as in the case of the homogeneous production function. The isoclines will be curves over the production surface and along each one of them the *K/L* ratio to simplify the statistical work. Homogeneity however is a special assumption. In some cases a very restrictive one. When the technology shows increasing or decreasing returns to scale it may or may not imply a homogeneous production function.

Law of Variable Proportions : Short-run Analysis

If one of the factors of the production (usually capital *K*) is kept constant, the marginal product of the variable factor (Labour *L*) will diminish after a certain range of production. The traditional theory of production concentrates on the range of output over which the marginal product of the factors are positive but diminishing. The ranges of increasing returns (to a factor) and the range of negative productivity are not equilibrium ranges of output. If production function is homogeneous with constant or decreasing returns to scale everywhere on the production surface, the productivity of the variable factor will necessarily be diminishing. If, however, the production function exhibits increasing returns to scale, the diminishing returns arising from the decreasing marginal product of the variable factor (labour) may be offset, if the returns to scale are considerable. This, however, is rare. In general the productivity of single-variable factor is diminishing.

Let us examine the law of variable proportions or the law of diminishing productivity (returns) in some detail.

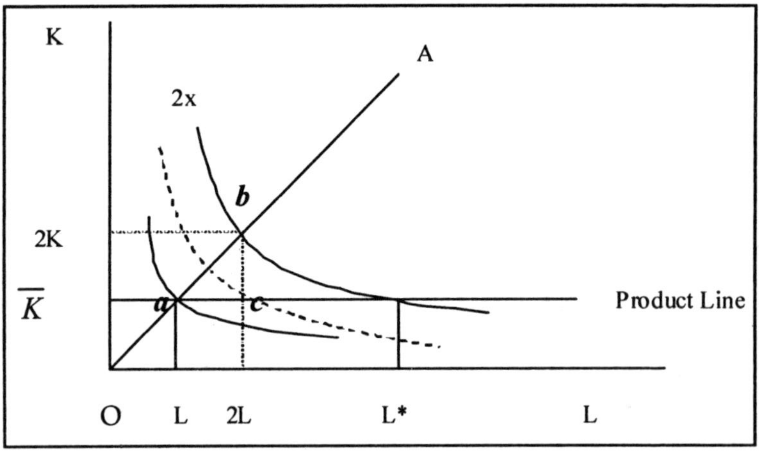

Fig. 4.13 : Law of Variable Proportions

If the production function is homogeneous with constant returns to scale everywhere, the returns to a single-variable factor will be diminishing. This is implied by the negative slope and the convexity of the isoquants. With *constant returns to scale* everywhere on the production surface, doubling both factors (*2K, 2L*) leads to a doubling of output. In Figure 4.13 point *b* on the isocline *OA* lays on the isoquant *2x*. However, if we keep *K* constant (at the level $\overline{K}$) and we double only the amount of *L*, we reach point *c*, which clearly lies on a lower isoquant than *2X*. If we wanted to double output with the initial capital $\overline{K}$, we would require *L** units of labour. Clearly *L* > 2L*. Hence doubling *L*, with *K* constant, less than doubles output. The variable factor *L* exhibits diminishing productivity (diminishing returns).

If the production function is homogeneous with decreasing returns to scale, the returns to a single—variable factor will be, *a fortiori*, diminishing. Since *returns to scale* are decreasing, doubling both factors will less than double output. In Figure 4.14(a) we see that with *2L* and *2K* output reaches the level *d* which is on a lower isoquant than *2x*. If we double only labour while keeping capital constant, output reaches the level *c*, which lies on a still lower isoquant.

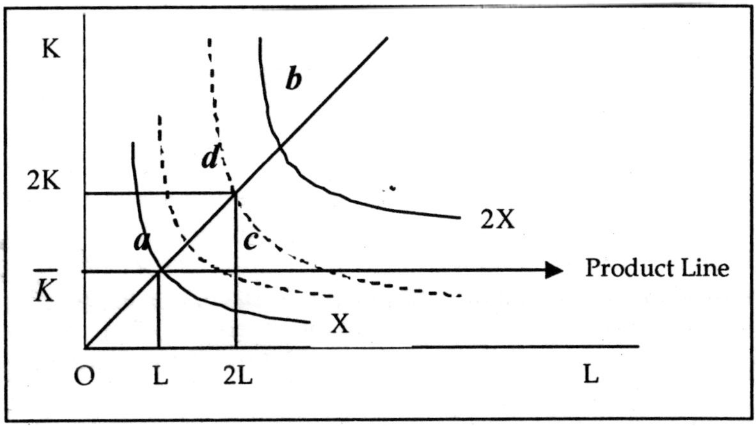

Fig. 4.14(a) : Law of Variable Proportions—Illustration (Decreasing)

If the production function shows increasing returns to scale, the returns to the single variable factor *L* will in general be diminishing [Figure 4.14(b)], unless the *positive returns to scale* are so strong as to offset the diminishing marginal productivity of the single variable factor. Figure 4.14(c) shows the rare case strong returns to scale which offset the diminishing productivity of *L*.

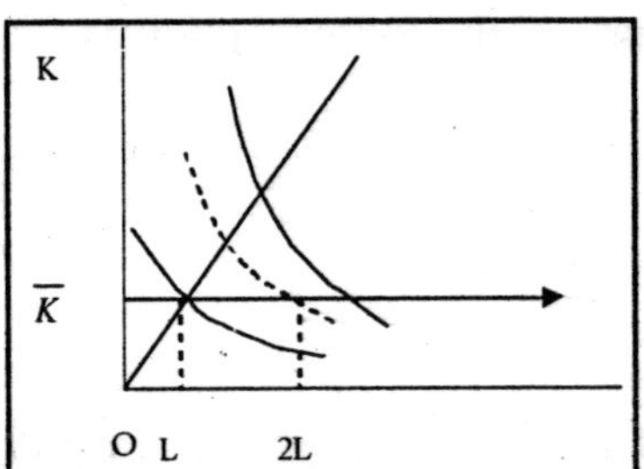

Fig. 4.14(b) : Illustration (Increase)

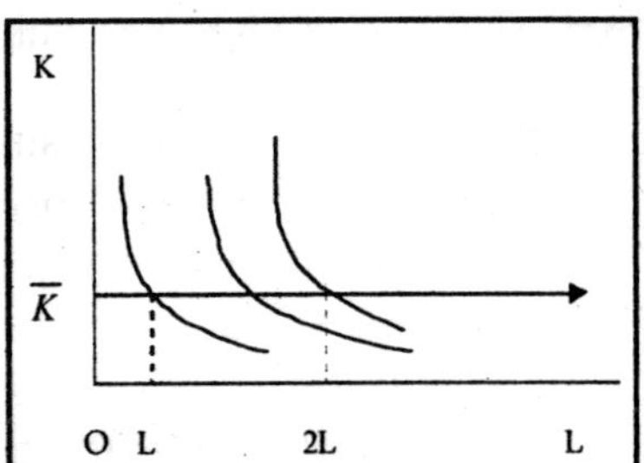

Fig. 4.14(c): Illustration

Optimal Combination of Factors of Production

In this section we shall show the use of the production function in the choice of the optimal combination of factors by the firm. In this part we will examine the case in which the firm is faced with a single decision, namely maximizing output for a given cost and minimizing cost subject to a given output. Both these decisions comprise cases of constrained profit maximization in a single period. It is assumed that the firm can choose the optimal combination of factors, that it can employ any amount of any factor in order to maximize its profits. This assumption is valid if the firm is new, or if the firm is in the long-run. However, an existing firm may be pressurized, due to pressure of demand, to expand its output in the short-run, when at least one factor, usually capital, is constant. The following assumptions are valid in the present analysis :

(i) The goal of the firm is profit maximization—that is, the maximization of the difference $\pi = R - C$, where π = profits, R = revenue, and C = cost;

(ii) The price of output is given, P_x;

(iii) The prices of factors are given: $\overline{w}$ is the given wage rate, $\overline{r}$ is the given price of capital services (rental price of machinery).

The problem facing the firm is that of constrained profit maximization, which may take one of the following forms :

(i) Maximize profit π, subject to a cost constraint. In this case total cost and prices are given (C, $\overline{w}$, $\overline{r}$, P_x), and the problem may be stated as follows :
$\text{Max } \pi = R - C$
$\pi = P_x X - C$
Clearly maximization of π is achieved in this case if cost C is minimized, given that X and P_x are given constants by the assumption.

(ii) Maximise profit π, for a given level of output. For example, a contractor wants to build a bridge (X is given) with the maximum profit. In this we have

$$\text{Max } \pi = R - C$$

$$\pi = \overline{P}_x \overline{X} - C$$

Clearly maximization of π is achieved in this case if cost C is minimized, given that X and P_x are given constants by assumption.

For a graphical presentation of the equilibrium of the firm (its profit maximizing position) we will use the isoquant map (4.15) and the isocost-line(s).

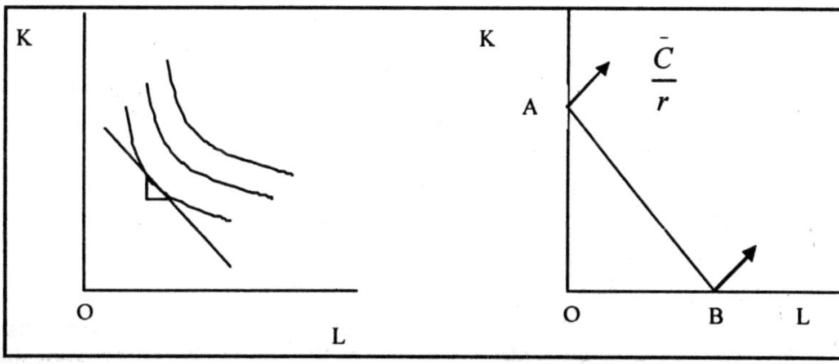

Fig. 4.15 : Isoquants *Vs* Isocost Line

The slope of an isoquant is

$$-\frac{\partial K}{\partial L} = MRS_{L,K} = \frac{MP_L}{MP_K} = \frac{\partial X/\partial L}{\partial X/\partial K}$$

The *isocost line is defined by the cost equation*

$$C\ (r)\ (K) + (w)\ (L)$$

Where w = wage rate, and r = price of capital services

The *isocost line* is the locus of all combinations of factors the firm can purchase with a given monetary cost outlay. The slope of the isocost line is equal to the ratio of the price of the factors of production. Slope of isocost line = w/r. In Figure 4.16 the maximum level of output the firm can produce, given the cost constraint, is X_2 defined by the tangency of the *isocost line,* and the highest *isoquant*. The optimal combination of factors of production is K_2 and L_2, for prices w and r. Higher levels of output (to the right of *e*) are desirable but not attainable due to the cost constraint.

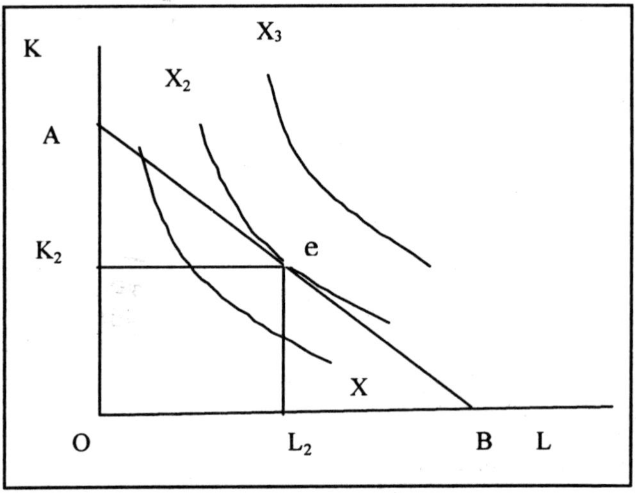

Fig. 4.16 : Optimal Combination

Other points on *ABO*, below it lie on a lower isoquant than X_2. Hence X_2 is the maximum output possible under the above assumptions (of given cost outlay, given production function, and given factor prices). At the point of tangency (*e*) the slope of the isocost line (*w/r*) is equal to the slope of the isoquant (MP_L/MP_K). This constitutes the first condition for equilibrium. The second condition is that the isoquants be convex to the origin. The conditions for equilibrium of the firm are :

(i) Slope of isoquant = Slope of isocost line;

$$\frac{w}{r} = \frac{MP_L}{MP_K} = \frac{\partial X/\partial L}{\partial X/\partial K} = MRS_{L,K}$$

(ii) The isoquants must be convex to the origin, if the isoquant is concave the point of tangency of the isocost curves does not define an equilibrium position (Fig. 4.17).

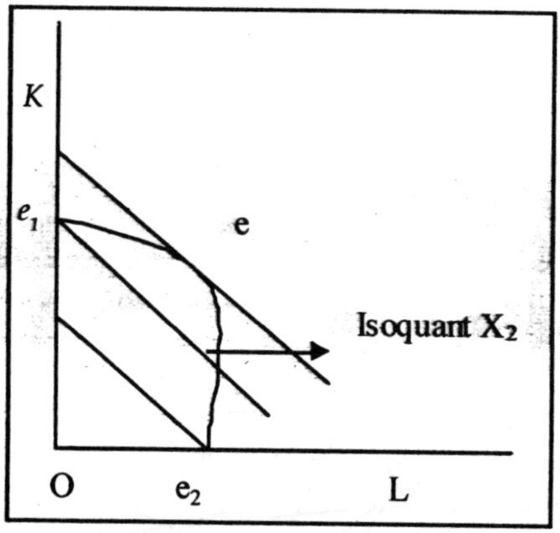

Fig. 4.17 : Concave Isoquant

Output X_2 (depicted by the concave isoquant) can be produced with lower cost at e_2 which lies on a lower isocost curve than *e*.

Optimal Expansion Path

In the long run all factors of production are variable. There is no limitation (technical or financial) to the expansion of output. The firm's objective is the choice of the optimal way of expanding its output, so as to maximize profits. With given factor prices (w, r) and given production function, the optimal expansion path is determined by the points of tangency of successive isocost lines and successive isoquants.

If the production function is homogeneous the expansion path will be straight line through the origin whose slope (which determines the optimal *K/L* ratio) depends on the ratio of the factor prices.

In Figure 4.18 the optimal expansion *OA* is defined by the locus of points of tangency of the isoquants with successive parallel isocost lines with a slope of w/r. If prices increases the isocost lines become flatter (for example, with slope of w'/r'), and the optimal expansion path will be the straight line *OB*. Of course, if the ratio of prices of factors was initially w/r and subsequently changes to w'/r', the expansion path changes :

initially the firm moves along *OA*, but after the change in the factor prices it moves along *OB*. If the production function is non-homogeneous the optimal expansion path will not be a straight line, even if the ratio of prices of factors remains constant. This is shown *w/r* ratio with $MRS_{L,K}$, which the same on a curved isocline.

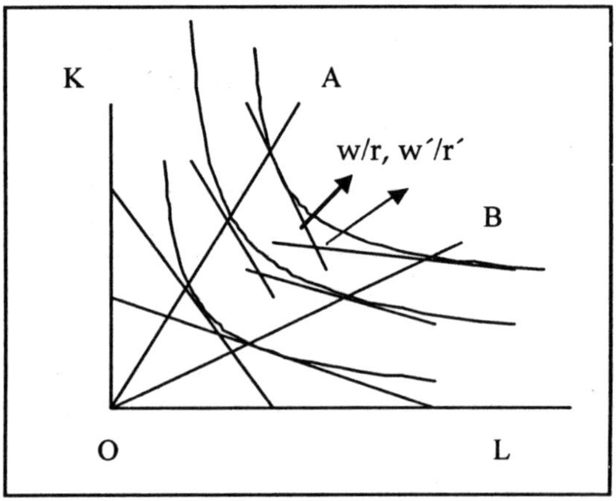

Fig. 4.18 : Optimal Expansion

In the short run, capital is constant and the firm is coerced to expand along a straight line parallel to the axis on which we measure the variable factor *L*. With prices of factors constant the firm does not maximize its profits in the short run, due to the constraint of the given capital.

Summary

One of the major elements of farm economics is agriculture production economics and is like any other sectors abide by the rules of basic economic theories. Production theory is based on production functions and laws of production. The production function is a purely technical relation which connects factor-inputs and outputs; it describes the laws of proportion, which is the transformation of factor inputs into products (outputs) at any particular time period. The production function includes all the technically efficient methods of production. Technology changes as knowledge of new and more efficient methods of production becomes available. In addition, new inventions may result in the increase of the efficiency of all methods of production. At the same time some techniques may become inefficient and drop out from the production function. These changes in technology constitute technological progress. This chapter also discussed about the laws of production. The laws of production describe the technically possible ways of increasing the level of production. Output can be increased by changing all factors of production especially in the long run. The *laws of returns to scale* refer to the long run analysis of production. On the other hand, in the short run output can be increased by using more of the variable factors, while fixed factors are kept constant. The expansion of output with at least one factor constant is described by the *law of diminishing returns* of the variable factor, which is often referred to as the *law of variable proportions*. The chapter concludes with optimal combination of factors of production and optimal way of expanding its output, so as to maximize profits. With given factor prices and given production function, the optimal expansion path is determined by the points of tangency of successive *isocost* lines and successive *isoquants*.

REFERENCES

Case, H.C.M. and Williams, D.B. 1957. *Fifty Years of Farm Management*. University of Illinois Press: Urbana.

Dillon, J.L. 1967. *The Analysis of Response in Crop and Livestock Production*. Pergamon Press.

Heady, E.O. and Dillon, J.L. 1961. *Agricultural Production Functions*. Iowa State University Press: Ames.

John M. Antle and Daniel A. Sumner. 1996. *The Economics of Agriculture: Selected Papers of D. Gale Johnson*. Vol. 1, University of Chicago Press, Chicago.

Laurens van der Laan. H., *et al.* (eds.). 1999. *Agricultural Marketing in Tropical Africa*. African Studies Center, Leiden.

Malcolm, L.R. 2000. *Proceedings of the 44th Annual Conference of the Australian Agricultural and Resource Economics Society*. Sydney.

Pannell, D.J. 1996. *Review of Agricultural Economics*, 18, 373-383.

5

Farm Production Decisions

Introduction

In this chapter we will apply the production theories that we discussed in the previous chapter. One distinguishing feature of agricultural production, as we discussed earlier, is that it always involves biological processes. This has many important implications for the application of economic theory to the study of agricultural production, marketing, and pricing. Agricultural production, crop production particularly, requires environmental inputs such as sunlight, water, heat, and time that are often beyond the control of farmers. The control of production, quality and yield is often far from complete. In addition, crop production requires land, so that the production activities are spread thinly over millions of acres.

Production activities involve the growth of animals and plants; the length of time between the application of inputs and the garnering of output might take anywhere from six weeks in the case of broilers to several years in the case of tree fruits. Crop production also typically takes place on an annual cycle. The result is that agricultural production involves heterogeneous outputs produced in small lots, often on a seasonal basis, scattered hither and thither over wide geographic areas.

Application of Production Function in Agriculture

The production function for any commodity is an equation, table, or graph showing the quantity of the commodity that can be produced per unit of time for each of a set of alternative inputs, when the best production techniques are used. A simple agricultural production function is obtained by using various alternative quantities of labour per unit of time to farm as fixed amount of land and recording the resulting alternative outputs of the commodity per unit of time.

Table 5.1 shows a short-run production function for wheat. Land is measured in hectares, labour in worker-years, and total products (*TP*) in tonnes per year. All units of land, labour or wheat are assumed to be homogeneous or of the same quality. The average product of labour (AP_L) is obtained by dividing each quantity of *TP* by the corresponding quantity of labour; land being the same. The marginal product of labour (MP_L) is obtained by the differences between the successive quantities of *TP*.

It is interesting for a farmer to know that wheat can be produced by combining land, seed wheat, nitrogen, phosphoric acid, potassium oxide, machinery, equipment and labour. However, this information would not be very useful unless the farmer know something about the manner in which these inputs must be combined in order to produce wheat. That is, he must know how and when to prepare the land, apply the fertilizer, and sow the seed. Otherwise it is possible to combine these inputs without getting wheat.

When a farmer is considering the question of how much nitrogen to be used in wheat production he may consider the other inputs as given or fixed in specified kind and quantities. In this case, we have:

$$Y_1 = f(X_1, X_2, X_3, \ldots\ldots\ldots, X_n)$$

That is, the amount of Y_1 (wheat) depends upon the amount of X_1 (nitrogen), given the amount of other inputs (X_2, X_3,………, X_n), which might be land, seed, wheat, labour, rainfall, temperature, etc. The vertical line indicates that input X_1 is variable and others are fixed in quantity.

Table 5.1 : Production Function of Wheat

Land	*Labour*	*TP*	AP_L	MP_L
1	0	0	0	—
1	1	3	3	3
1	2	8	4	5
1	3	12	4	4
1	4	15	3.75	3
1	5	17	3.40	2
1	6	17	2.83	0
1	7	16	2.28	–1
1	8	13	1.63	–3

Box 5.1: Illustration of Production Function

Mr. Prasad, a farmer in Tamil Nadu cultivates Wheat in 10 ha of fertile land. The productivity of Wheat is 1765 kg/ha without adding fertilizer. The scientists of Agriculture University indicate that productivity of Wheat will increase at 1.5 times of every dose of one unit of NPK fertilizer. And, finally, Mr. Prasad applied 100 kg of fertilizer per ha. Calculate the following :

(a) Production function of Wheat,
(b) Production of Wheat without applying fertilizer,
(c) Production of Wheat after applying fertilizer.

Solution

(a) $Y_1 = a + bX_1$
$Y_1 = 1765 + 1.5\,X_1$
(b) Production of Wheat without applying fertilizer = 1765/ha
Production in 10 ha = 1765 x 10 = 17650 kg
(c) Production of Wheat after applying fertilizer (Y_1) = 1765 + 1.5 X_1
= 1765 + 1.5 x 100 = 1765 + 150
= 1915 kg
Production in 10 ha = 1915 x 10 = 19150 kg

The manner in which the amount of wheat varies as the amount of nitrogen varies depends upon the amount of nitrogen used. It is possible that there is a constant relationship between application of nitrogen and the amount of wheat produced. In this case the production function may be in the following form:

$$Y_1 = a + bX_1$$

For example, if farmer expand one unit of wheat for each two units of nitrogen applied, then the linear relationship is as follows:

$$Y_1 = a + 0.5\,X_1$$

Where, *a* indicates the yield of wheat without applying nitrogen, and 0.5 indicates that Y_1 increases by 0.5 unit for every increase of one unit of nitrogen.

Cobb-Douglas Production Function

The Cobb-Douglas production function is the most widely used production function in empirical works. The function is expressed by

$$Q = AL^{\alpha}K^{\beta}$$

Where, *Q* is output and *L* and *K* are inputs of labour and capital respectively. A, α, and β are positive parameters determined in each case by the data. The greater the value of *A*, the more advanced is the technology. The parameter α measures the percentage increase in *Q* resulting from a 1% increase in *L*, while holding *K* constant. Similarly, β measures the percentage increase in *Q* resulting from a 1% increase in *K* while holding *L* constant. Thus, α and β are the output elasticity of *L* and *K* respectively. If $\alpha + \beta = 1$, there are constant returns to scale, if $\alpha + \beta > 1$, there are increasing returns to scale, and if $\alpha + \beta < 1$, there are decreasing returns to scale.

Irrational Production in Agriculture

By using the data of Table 5.1, we can show the *TP, AP* and *MP* as follows :

The input-output relation showing *TP, AP,* and *MP* can be divided in three phases in such a manner that we can isolate the portion of the production function in which production is most profitable. In Figure 5.1, Phase I of the production function goes to the level of input at which the maximum *AP* is obtained. In this phase, the *AP* is increasing and *MP* is greater than *AP*. Here it will always be profitable for the producer to continue to add inputs as long as the *AP* is increasing. For example, in a situation in which one input, say land, is free and another input, say labour, is expensive and limited. Under this situation it pays to limit the quantity of land used in relation to the quantity of labour used, even though land is free.

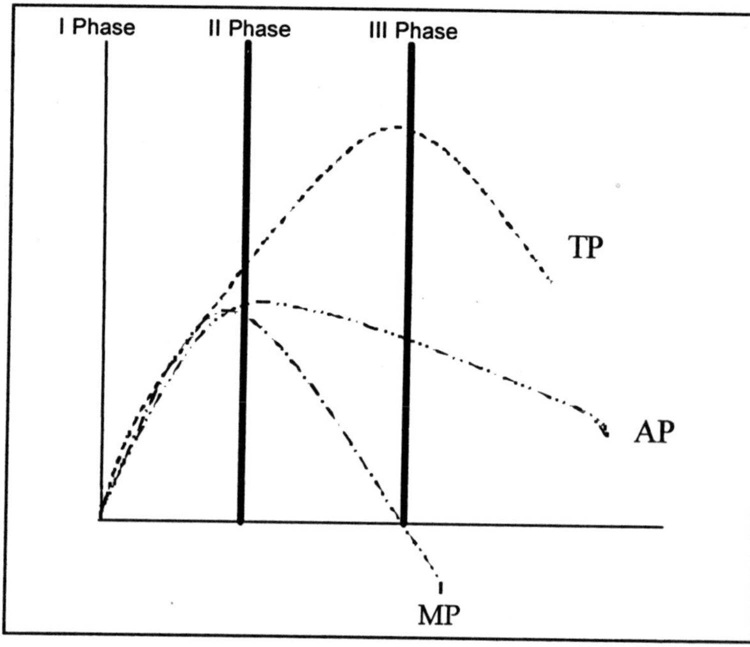

Fig. 5.1 : Production Function

In the III phase, the total product is decreasing. Hence, in the III phase, the *MP* or the amount of product added by additional units of input is negative. Since additional quantities of input reduce total output in the III phase, it is not profitable to operate with a combination of resources existing in this phase. The point at which *MP* becomes zero represents the maximum quantity of variable input which will ever be profitable to use in combination with other inputs. For example, if labour are free and land expensive, it would be profitable to add labour up to the point, when the return from land is at the maximum. At this point, the *MP* of an additional unit of labour is zero. If labour is added beyond this point, the *MP* is negative, and it does not pay to operate in this region.

Thus, the Phase-I and Phase-III set the limits to the phase in which it is profitable to operate. Therefore, Phase-II is the region in which production is most profitable. In Phase-II, *TP* is increasing; *MP* is decreasing, is positive, and is less than the

AP; and the *AP* is decreasing. This is the phase of rational production. But at the same time it should be emphasized that this region is one of diminishing returns; both *AP* and *MP* are decreasing.

Box 5.2 : Illustration of Cobb-Douglas Production Function

Values of α and β for different agricultural products in selected countries are given in the following table.

Crops	*Country*	α	β
1. Teff	Ethiopia	0.70	0.41
2. Barley	France	0.83	0.10
3. Wheat	Japan	0.80	0.37
4. Maize	India	0.20	0.67
5. Sorghum	United States	0.71	0.29
6. Millet	Germany	0.72	0.35
7. Oats	Russia	0.80	0.38

Calculate the following :

(a) Derive the Cobb-Douglas Production function for each crop in the corresponding countries.

(b) What type of returns to scale prevails in each crop?

(c) How much does output rises in the US sorghum production if *L* increases by 1 per cent?

Solution

(a) Cobb-Douglas production function and (b) Returns to scale

Crops	*Country*	Production function	Returns to scale
1. Teff	Ethiopia	$Q = AL^{0.70}K^{0.41}$	Increasing
2. Barley	France	$Q = AL^{0.83}K^{0.10}$	Decreasing
3. Wheat	Japan	$Q = AL^{0.80}K^{0.37}$	Increasing
4. Maize	India	$Q = AL^{0.20}K^{0.67}$	Decreasing
5. Sorghum	United States	$Q = AL^{0.71}K^{0.29}$	Constant
6. Millet	Germany	$Q = AL^{0.72}K^{0.35}$	Increasing
7. Oats	Russia	$Q = AL^{0.80}K^{0.38}$	Increasing

(c) Since $\alpha = 0.71$ in the US sorghum production, a 1 percentage increase in *L* only would increase *Q* by 0.71 per cent.

In agriculture, producers do not have perfect knowledge of production relations. Because of this imperfect knowledge, it is possible to observe that producers combine resources in such proportions that they operate in Phase-I or III of the production function. For example, in the production of many crops insufficient time is used to reach Phase-II. Adding more time would increase the *AP* from use of the time.

On the other hand, evidence of production in Phase-III is often noted. For example, during the later summer, we can see in rural India too many cattle on a given quantity of pasture, resulting in over-grazing of pastures and less production than could have been obtained with fewer cattle.

Application of Input-Output Analysis

Economic decisions by farmers frequently fall in the category of problems concerning the most profitable level of application of one input, given the level of other inputs, in the production of one commodity.

Under many soil and climatic conditions, the amount of agricultural commodity a farmer can expect to produce depends on the amount of inputs including fertilizers used per unit of land. In determining how much fertilizer to apply, the farmer must compare the added revenue which would be received from the application of fertilizer with the additional cost. He estimates the additional revenue expects to receive by multiplying the additional amount of the agricultural commodity expected to be harvested by the price he might reasonably expect to receive for his agro-products. Additional costs include expenditures for additional fertilizer, expenses involved in the application of the fertilizer, and expenses involved in harvesting and marketing the additional agro-products. If the farmer is to make decision consisted with maximization of net revenue, it is necessary for him to know how each of these items varies with the level of application of fertilizer.

The amount of fertilizer application varied in different fields depending upon the inputs, including weather. Experiments in US shows, under dry weather and wet weather conditions, diminishing returns are in evidence (Figure 5.2). The difference in dry weather yield and good weather yield increases as the law of fertilizer application increases.

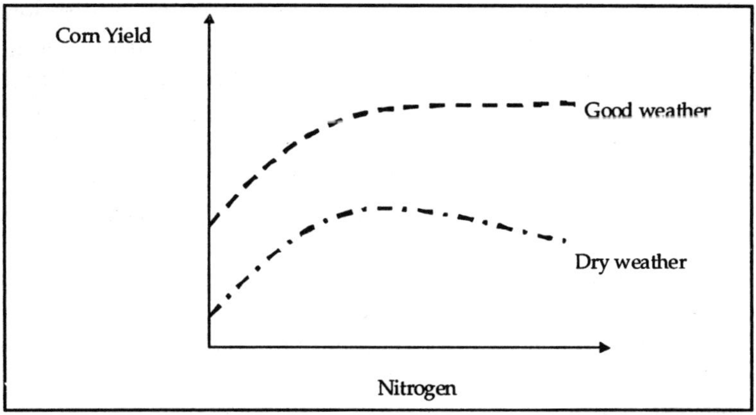

Fig. 5.2 : Response of Corn to Nitrogen, USA

Weather conditions are important in determining the increase in yield resulting from the application of fertilizer. However, the farmer cannot control weather conditions.

The quantity of an input most profitable to use in the production of a commodity is expected to change as the price of the input and the price of the commodity changes. The extent to which the most profitable level of application of an input varies depends on the shape of the production function and on the price of the input in relation to the price of the product.

Application of Cost Functions

Cost analyses are often helpful to farmers and processors in making decisions. Among other uses, cost analysis can be used to determine whether an enterprise is profitable, what size of the plant should be constructed, and what level of output is necessary before it pays to purchase some machine.

The low price of grains at harvest and the higher price later provide an incentive for grain producers to consider the possibility of increasing their net revenue by storing grains for sale at the later date. However, costs are associated with storage of grain. It may even be necessary to dry the grain artificially before it can be safely stored. Such variable costs include expenditures for cleaning bins, labour, hauling to bins, electricity, insect control, fuel for grain moving equipment, grain insurance, and loss in weight or shrinkage during storage. A farmer contemplating storage, and who does not already own storage facilities, must consider all costs, which include the costs of adding such facilities. To such farmers, all costs of storage are variable. He must include as variable costs the cost of interest on investment, property taxes, annual repairs, insurance and depreciation.

The volume of grains which is dried and stored has a large effect on the total costs. Therefore, the volume stored affects the farmer's decision as to whether to build drying and storage facilities and store grain on his farm or to sell at the time of harvest.

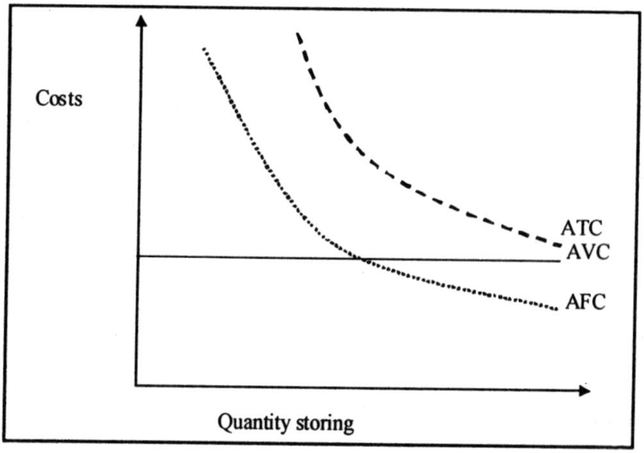

Fig. 5.3 : Costs of Storage

Farmers have an additional alternative in storing grain. They can store grain in commercial facilities. When this is possible, a farmer may find that commercial storage is most profitable for small volumes, whereas on-farm storage result in more net revenue if the farmer has large volumes to store. A farmer may find that commercial storage is cheaper than on-farm storage for any volume stored unless on-farm facilities are already available. In any case, all available alternatives must be weighed by a farmer before making his decision.

Family Farm Model

The economic agents visualized in economic theory have limited affinity with their real-world counterparts. To make theory conform more realistically to agriculture, we define a family farm as a farm in which the family or household supplies labour, equity capital, and management.

The following are the major assumptions of the family farm model :

(i) No distraction between managerial labour and labour used to plant crops, feed animals, milk cows, etc.
(ii) The authority for decision may be diffused throughout the entire family.
(iii) Once a household organizes a farm (enters farming) all the family labour is fixed in farming.
(iv) The family farm is free to hire labour; hired labour and family labour are not perfect substitutes.

The definition of profit in short-run model is as follows:

$$\pi = TR - TVC - TFC \quad \text{.......... (1)}$$

Where, π is profit, *TR*—total revenue, *TVC*—total variable cost, and *TFC*—total fixed cost. Total revenue is calculated by multiplying price times output. Total variable cost equals the sum of the products of variable input quantities times their respective prices.

$$QR = TR - TVC \quad \text{.......... (2)}$$

Where, *QR* is the quasi-rent. Since quasi-rent is the net return over variable cost; it can be interpreted as the amount of money available to remunerate the fixed factors, that is, as the potential return to fixed equity capital and fixed family labour.

Equation (1) can be written as:

$$\pi = QR - TFC \quad \text{.......... (3)}$$

TFC is break up into components, one attributed to fixed capital and the other to family labour. The fixed cost of capital is denoted by *CC* (cost of capital). The fixed cost of family labour is denoted by *TE* (transfer earnings).

A household entering farming means the family labour will be fixed in the farm business. Taking family labour as the 'residential claimant', the return to family labour (*RFL*), is defined as:

$$RFL = TR - TVC - CC \quad \text{.......... (4)}$$

Household's transfer earnings (*TE*) is defined as the minimum *RFL* that would prompt the household to enter farming. For a household that is already established in farming, *TE* is the minimum *RFL* that would keep them in farming.

The household decision regarding entering farming is as follows :

(a) If $RFL > TE$, the household enters farming.
(b) If $RFL < TE$, the household does not enter farming.
(c) If $RFL = TE$, the household is indifferent between farming and next best alternative career. Here the farmer is said to be a marginal farmer.

That is, profit is given by :

$$\pi = TR - TVC - CC - TE \quad \text{.......... (5)}$$

$$\pi = TR - TVC - TFC \quad \text{.......... (6)}$$

Where, *TFC* equals the sum of *CC* and *TE*. Now the farmers' choices regarding entering the farming are as follows :

(a) If $\pi > 0$, the household enters the farm industry. The returns to the family's resources in farming exceed the return in the non-farm sector.
(b) If $\pi < 0$, the household stays out of farming.

Although *CC* is based on opportunity cost of capital, *TE* is not based on the opportunity money cost of labour. Rather, *TE* is based on opportunity cost of labour when both money and work satisfaction are taken into account.

Box 5.3 : Definition and Illustration of Cost of Capital

If a household has $ 100 at the beginning of the current year (Year-1), but owns no other assets. The household plans to retire at the end of some future year, year-*T* (Say 20 years). The $ 100 could be invested in a money market fund at an annual rate of interest given by *i* (Say 5%). How much money would be available at the end of year T?

The formula is: $M_T = M_0 (1+i)^T$ .... (A)

Where, M_0 is the initial amount of money invested at the beginning of year1 and M_T is the amount to be withdrawn at the end of year *T*.

Solution

$$
\begin{aligned}
M_0 &= \$\,100 \\
i &= 5\% = 0.05 \\
T &= 20 \\
M_{20} &= 100\,(1 + 0.05)^{20} \\
&= \$\,265.33
\end{aligned}
$$

If the household has the alternative of purchasing a farm complete with land and machinery. The farm would be operated entirely by hired labour. At the end of each year, the household receives quasi-rent (*QR*), which is deposited in the money market fund. How much money will have accumulated in the money market fund under the investment option at the end of year *T*?

Let QR_t the quasi-rent to be invested in the money market fund at the end of each year *t* (t = 1, 2, 3,........*T*) and will be constant for all *t*. At the end of period *T*, the household would have:

$$QR\,(1 + i)^{T-1} + QR\,(1 + i)^{T-2} + QR\,(1 + i)^{T-3} + \ldots\ldots\ldots\ldots +$$

$$QR\,(1 + i)^{T-T} + S = QR \sum_{T-1}^{T} (1 + i)^{T-t} + S \ldots\ldots\ldots\ldots\ldots\ldots\ldots \text{(B)}$$

Where, *S* is the salvage value or the price of the farm assets at the end of year *T*.

$$\text{That is, } M_T = CC \sum_{T-1}^{T} (1+ i)^{T-t} + S \ldots\ldots\ldots\ldots\ldots\ldots\ldots \text{(C)}$$

From equation *c*, we get

$$CC = \frac{M_I - S}{\sum_{T-1}^{T}(1+i)^{T-t}} \ldots\ldots\ldots\ldots\ldots\ldots\ldots \text{(D)}$$

Equation-D defines CC, the opportunity cost of capital tied up in the fixed plant. In general, CC equals user cost of capital. That is, the implicit valuation of the flow of services provided by fixed factors over many years. In real world situation, it reflects the interest rate, depreciation, and other factors as well. Aside from considerations of risk, an entrepreneur would buy the plant if the resulting wealth at the end of year T were larger than M_T

Figure 5.4 shows the short-run cost curves for a family farm; giving curves for average fixed cost (AFC), average variable cost (AVC), average total cost (ATC), and marginal cost (MC). Fixed cost is divided into two components: average cost of capital (ACC), defined as CC divided by output, and average transfer earnings (ATE) defined as TE divided by output. The AFC curve is found by summing vertically ACC and ATE. ATC is found by summing vertically AVC and AFC.

In Figure 5.4, at 'Po' price total revenue (Poqo) equals the sum of total variable cost (AVCoqo), cost of capital (ACCoqo), and the transfer earnings (ATEoqo). Profit equals zero and the returns to family labour (RFL) equal TE.

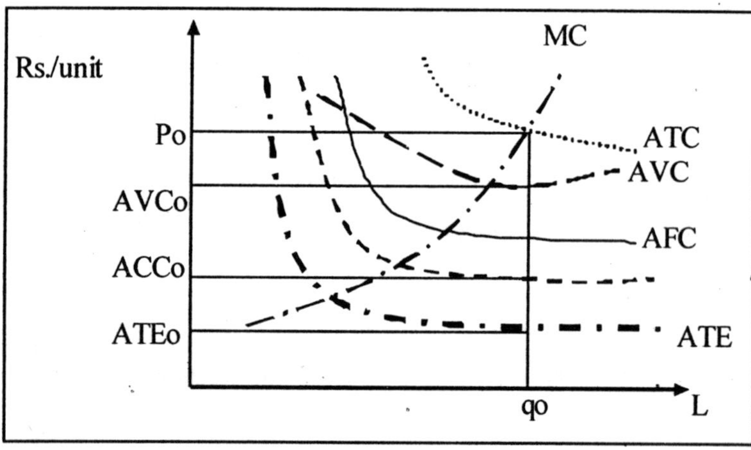

Fig. 5.4 : Cost Curves of Family Labour

The supply function of farm operations as a function that shows how many households would be willing in the long run to organize and operate farms (enter or stay in farming) at alternative level of returns to family labour (RFL). If transfer earnings equal TEo for all households, then the farm operator's supply function is perfectly elastic. If the supply function for all farm inputs, including that for farm operators are perfectly elastic, then the long run supply function for farm output will also be elastic.

Farm Output and Input Pricing

Consider a farm choosing land *a* and other inputs *k* in the production of output. For simplicity, we will consider only the single-output case. Let *p* denote the expected output price, *J* be the price of input *k*, and *R* be the rental price of land. Assume that output and output price are independently distributed random variables and that the farmer is risk neutral. Then the farmer makes decisions by maximizing expected profit.

$$\pi(p, J, R, TFC) = Max_{k,a}\ [pf(k,a) - J'k - Ra - TFC] \text{ ....(7)}$$

Where, *f(k,a)* denotes expected output, *pf* is expected revenue, *Ra* is the rental cost of land. J'K is the cost of inputs other than land, and TFC denotes total fixed cost.

Assuming that the farm chooses to produce and to use positive amounts of inputs (that is, quasi-rent is not negative), the first-order conditions associated with the maximization problem are:

$$P\frac{\partial f}{\partial k} = J \text{..........(8a)}$$

$$P\frac{\partial f}{\partial a} = R \text{..........(8b)}$$

The expression 8 states that, at the optimum for a producing farmer, the expected marginal value product (left hand side of the equation-8) equals the input cost (right-hand side of equation-8).

Let *k'(p, J, R)* and *a'(p, J, R)* denote the optimal input choices for *k* and *a* corresponding to equations 7 and 8.

The optimal supply function is given by

$$q'(p, J, R) = f(k', a')$$

Equation-7 is rewritten as:

$$\pi(p, J, R, TFC) = Max_q\ [pq - C(J, R, q) - TFC] \text{ ....(9a)}$$

where, *q* denotes expected output and

$$C(J, R, q) = Min_{q,a}\ [J'k + Ra : q = f(k,a)] \text{ ———————— (9b)}$$

is the cost function characterizing the input choices that minimize the cost of producing a given expected output level *q*.

Denoted by k^c (J,R,q) and a^c (J,R,q) the corresponding cost minimizing input demand function. The equivalence of equation-7 and 9 implies the following:

$k'(p, J, R) = k^c [J, R, q'(p, J, R)]$ ....(10a) and
$a'(p, J, R) = a^c [J, R, q'(p, J, R)]$ ....(10b)

Expression-10 indicates that the cost minimization problem (9b) is fully consistent with the profit maximization problem (equation-7).

Assume that the land is necessary input such that $f(k,0) = 0$ for any $k \geq 0$. Define quasi-rent (ϕ) as total revenue (pq) minus total variable cost [$C(J, R, q)$]. Then the farmer would decide to produce only if $\phi(p,J,R) > 0$. Indeed if $\phi(p,J,R) < 0$, then the farmer would be better off producing nothing. In this case $\phi = 0$ and $p = -TFC$.

Analysis of Farm Programmes: Direct Payment Programme

In a direct payment programme, the government announces a target price p' for a farm commodity, say food, and makes direct payments to farmers to assure that the real prices they receive will not fall below p'. That is, the farmer's output is sold at the market price p. If the market price equals or exceeds the target price p', the government does nothing. If the market price falls below p', the government sends a check to the farmer equalling $(p' - p)\, q$, where 'q' equals the farmer's output. The direct payment per unit of output equals $(p' - p)$.

In real world applications, it is important that the direct payment be based on the market price, an average for the nation, say, as opposed to the actual search for the highest price.

In Figure 5.5, DD is demand curve and SS supply curve. Intersection of the curves give the competitive price *Pc* and the competitive output *Qc*. For any market price less than the target price *Po'*, farmers receive a per unit direct payment that brings the price they effectively receive up to *Po'*. For all such market prices the effective farm price equals *Po'*, and in the aggregate, farmers produce *Qg*. For market prices in excess of *Po'*, the market price becomes the effective price. Thus the supply curve that shows how much farmers would be willing to produce at alternative market prices is given by the curve *SSg,* kinked at *Po′* (Point *a*). The new equilibrium price *Pg* and output *Qg* are given by the intersection of *DD* and *SSg.*

In equilibrium buyers are able to buy what they want at the marker price *Pg;* farmers sell as much as desired at the effective farm price *Po′*. Market participants are in equilibrium because they could not further their goals by changing their decisions. The farmer receives a government check equal to *(Po′ – Pg)q,* with the direct payment per unit equalling *(Po′ – Pg).* Government expenditure equals *(P0′ – Pg) Qg.*

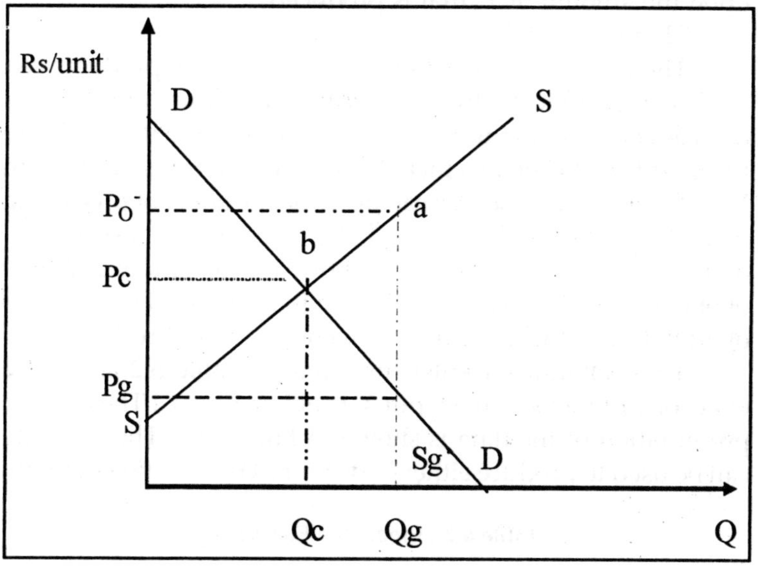

Fig. 5.5 : Equilibrium under Direct Payment Programme

Product Choices and Production Possibilities

A common problem faced by the farmers or producers is to decide how to allocate the resources he owns, rents, or borrows to the alternative products, which can be produced. It involves choices between the number of cows and sheep, the acres of rice and wheat, or coffee and tea, etc. A producer wishes to obtain maximum satisfaction from the use of his resources. Farmers usually wish to know how much revenue will have to be given up to have some combination of enterprises other than that which would give maximum net revenue. When a farmer states that he prefer chickens to cows, it means that, if chicken yield the same or greater net revenue from his resources than cows, he will choose chickens.

A producer is interested in the effect that producing more of one product has on the production of alternative products. For example, he is producing two products, Y_1 and Y_2 using the same resource X_1. The amount of X_1 that can be used to produce Y_1 depends upon the amount of X_1 used in producing Y_2. Therefore, the amount of Y_1 that can be produced depends upon the amount of Y_2 that is produced.

That is, $Y_1 = f(Y_2)$

The relationship existing between any two products can be shown graphically by considering the amount of resources used as constant. In this way it can be shown, what happens to the quantity of each product as resources are transferred from Y_1 to Y_2, or the reverse. We can describe the effect of producing more Y_1 on the production of Y_2 by 'production possibilities' or 'iso-resource relationships'. It shows the maximum amounts of one product (Y_1) which can be produced for particular levels of another product (Y_2) from a given quantity of resources.

Two separate product functions in Table 5.2 show the effect of adding the same input X_1 to either Y_1 or Y_2. The graphic presentation of the data is shown in Figure 5.6. The input X_1 can be used to produce either Y_1 or Y_2 or a combination of them.

Table 5.2 : Production Functions

Input (X_1)	*Output (Y_1)*	*Output (Y_2)*
0	0	0
5	7	11
10	13	20
15	18	28
20	22	35
25	25	41
30	27	46

If only 30 units of X_1 are available to the producer for production of Y_1 and Y_2, we can determine various combinations of the two products in such a manner that a maximum of one is obtained for any level of the other (Table 5.2). If the producer uses all 30 units of X_1 in the production of Y_2, he obtains 46 units of Y_2 but does not get any Y_1. He can shift 5 units of X_1 to the production of Y_1. In doing so, he receives 7 units of Y_1, but only 25 units of X_1 are available to produce Y_2. He can produce only 41 units of Y_2 with these 25 units of X_1.

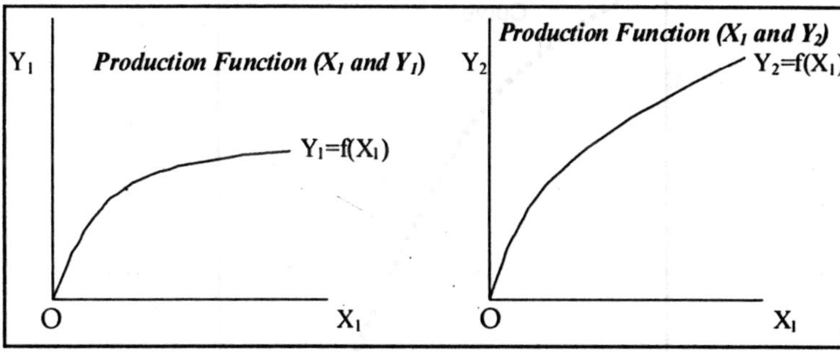

Fig. 5.6 : Production Function

The data are shown as production possibilities curve in Figure 5.7. Commodity Y_1 on the X-axis and Y_2 on the Y-axis. The way in which two products can be substituted for each other is important from an economic stand point. There are as many production possibilities curves as there are input levels.

Table 5.3 : Possible Product Combination for 30 Units of Input

Units of X_1 used in the production of		*Output*		*Marginal Rate of Substitution*
Y_1	Y_2	Y_1	Y_2	$(\Delta Y_2/\Delta Y_1)$
0	30	0	46	—
5	25	7	41	–5/7
10	20	13	35	–6/6
15	15	18	28	–7/5
20	10	22	20	–8/4
25	5	25	11	–9/3
30	0	27	0	–11/2

The maximum net revenue obtainable with a given amount of inputs is where the physical rate of substitution between two products is equal to the rate at which the products exchange in the market, or where $\Delta Y_2/\Delta Y_1 = P_{Y1}/P_{Y2}$

If, as inputs are transferred from Y_2 to Y_1, $P_{Y1}\Delta Y_1 > P_{Y2}\Delta Y_2$, more inputs should be added to produce Y_1 and less Y_2 should be produced. On the other hand, if $P_{Y1}\Delta Y_1 < P_{Y2}\Delta Y_2$, adding inputs to the production of Y_2 will be profitable.

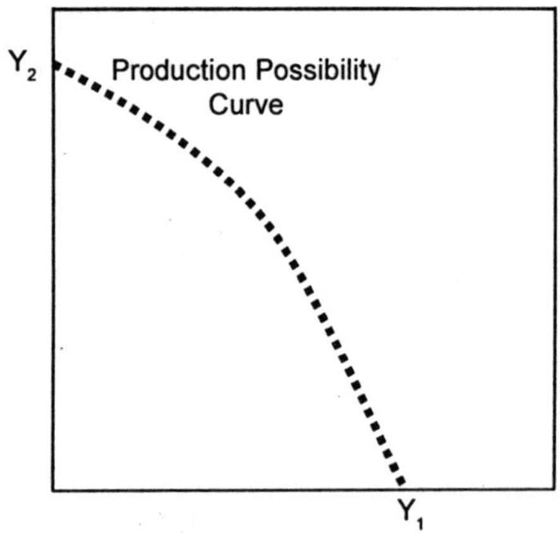

Fig. 5.7: Production Possibility

Farm Organization and Budgets

Net revenues farmers getting vary widely from farm to farm. Differences in net revenues of farm families result from differences in kinds and quantities of resources, which they own and differences in the use which they make of these resources. Farmers can compare the costs of different methods of performing operations or they can compare net revenues from producing different kinds of products. In order to make these comparisons, farmers must have information in regard to the consequences of alternative choices. Construction of enterprise budgets is one method by which producer can compare directly the revenues and costs of alternative enterprises.

A budget is a plan or a record of what one might expect to obtain from particular production practices. An enterprise budget consists of a statement of expected revenues and expenses from the products of a particular product. Development of a budget requires information pertaining to the amount of product which should be expected from particular quantities of inputs and information in regard to the expected prices of the product and of the inputs.

We can express the enterprise budget in the following equation form :

If $Y_1 = f(X_1, X_2, \ldots\ldots\ldots, X_n)$

Then Net Revenue $= Y_1P_{Y1} - (X_1P_{X1} + X_2P_{X2} + \ldots\ldots\ldots + X_nP_{Yn})$

An enterprise budget is shown in Table 5.4 for cotton production. It shows the amount of cotton which might be expected over the long-run from particular levels of application of fertilizers and other inputs.

Table 5.4 : Sample Enterprise Budget for Cotton Production (1 Acre)

Particulars	*Price (Rs.)*	*Units*	*Quantity*	*Amount (Rs.)*
I. IncomeYield				
a. Lint	0.315	Pounds	625	196.88
b. Seed	0.031	Pounds	1060	32.86
Total				229.74
II. Expenses				
a. Seed	2.75	Bushels	1	2.75
b. Fertilizers	0.275	Pounds	190	15.25
c. Pesticides	0.26	Pounds	60	15.60
d. others (Labour)	0.05	Hours	45.5	58.66
Total				92.26
Net Revenue				229.74 – 92.26 = 137.48

The items which should be included in an enterprise budget are those variable inputs which are used in the production of the particular commodity under consideration. In the enterprise budget shown in Table 5.4, only the variable expenses are charged to the cotton enterprise. No charge was made for the fixed expenses of land, equipment, and family labour. On the same manner, if a farmer has more than one crop on the farm, he would not charge all of the combined expenses to one crop. Such expenses are put in the category of fixed cost and are not relevant in influencing the farmer's production decisions as to choice of enterprises.

A farmer can construct budgets for each enterprise he wishes to consider on his farm. He needs to obtain information as to the amount of inputs and information as to input and output prices. A farmer can tell from the construction of a budget whether he could obtain positive net revenue from the production of a commodity.

Land Tenure Arrangements

Land tenure refers to the possession of rights to the use of land. People hold varying kinds of rights in the use of land and are said to belong in different tenure classes. Although it is difficult to rank tenure classes according to the degree of rights which are held, it is generally recognized that the owner-operator without debt has the most freedom of action with respect to the use of his inputs. At the other end of this scale of rights in land are found the hired farm labourers and share-cropper's. Between these two extremes are share tenants, cash tenants, mortgaged owners, part-owners, and numerous combinations of these groupings.

There are many different types of tenure arrangements in agriculture economy world-wide. These arrangements influence the efficiency with which inputs are used. They also affect the degree of uncertainty encountered in the operation of a farm. Although innumerable breakdowns are possible, most tenure arrangements can be placed in one of the following :

(i) *Owners :* This group includes those who have title to all of the land they operate. This is known as ownership in fee simple. It entitles the owner to sell or use his land as he sees fit within the limits of the law. Owners as a group generally have more freedom in their production plans than any other tenure class. However, in order to obtain ownership many owners have had to be satisfied with relatively small farms because of the small amount of capital available to them. Many farmers are capable of managing larger farms than they are able to buy. Also, where there owners have a large debt, they may have to adjust their production so that their short-term income will be sufficient to meet living expenses plus payments on their debt obligations. This may mean that a farming pattern will be necessary which will not yield the greatest long-term income for the family. For this reason, we may find that ownership does not always guarantee the most efficient organization of resources.

(ii) *Part-Owner* : This group includes those farmers who own some land and rent additional land in order to enlarge their farming units. Part-owners have, on the average, large farms in terms of acres and in value of land and building than do owner-operators or tenants. Renting additional land is a way of gaining control of larger quantities of resources without increasing risk by going further into debt. With the relatively high capital requirements needed for efficient farming, many farmers have found this form of tenure to be very satisfactory.

(iii) *Tenants* : Tenants are those farmers who rent all of the land they operate. Several different tenant categories exist. Share-croppers, for example, are tenants who pay a particular type of rent, a share of the crop. The basic function of tenancy is to allow two or more people to combine their inputs into a single producing unit. Unfortunately, conflicts between land lord and tenant can arise with leasing arrangements.

Supply of Individual Farm Products

Production of agricultural products varies substantially over time. Some of the changes in quantities of commodities produced likely result from changes in uncontrollable conditions such as weather. On the other hand, many changes in production result from changes in relative prices and costs. In some years there are fairly definite cycles of production of agricultural products. A possible explanation of the cyclical behaviour is demonstrated by the 'Cobweb' theorem of production.

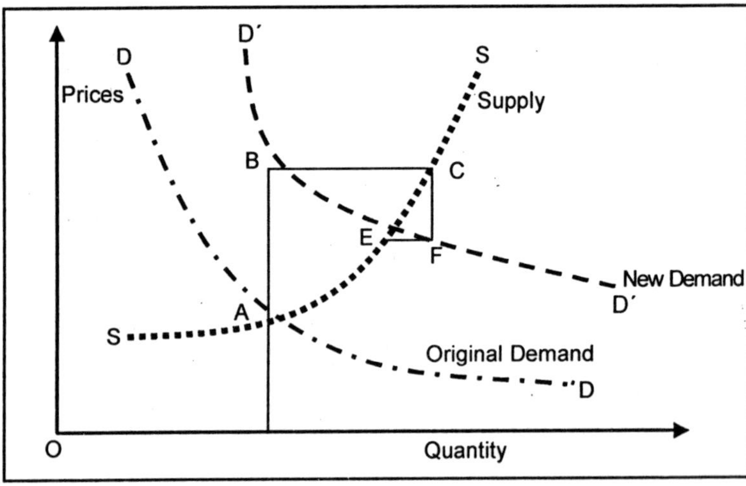

Fig. 5.8 : Cobweb behaviour in agriculture production

According to the Cobweb theorem, producers of a commodity expect the price in the next period of production to be the same as the current price and make production plans accordingly. As shown in Figure 5.8, the industry is in equilibrium at *A*, that is, the consumers are willing to purchase the same quantity that the producers are willing to supply. The original demand curve *DD* shows the maximum quantities of the good that consumers are willing to purchase at alternative prices. If the demand shifts *D′D′* because people's preferences change so that they are willing to buy more of the goods at all prices. Now the consumers are willing to pay the price at *B* for the same output for which they originally were willing to pay the price at *A*. In planning their production, producers consider the current price at *B* as the future price and make plans to increase their production to *C* accordingly. When production has been increased, however, it is necessary to take a lower price for the product than was anticipated. Consumers are willing to pay a price equal to that at *F* for the new output. In view of this, producers plan to decrease their production in the next time period to the amount at *E*. In some cases, this process continues until an equilibrium level of output at the intersection of the supply curve and the new demand curve is reached. In the other case, equilibrium is never reached, and has a continuous fluctuation in prices and production.

Summary

One distinguishing feature of agricultural production is that it always involves biological processes and it has many important implications for the application of economic theory to the study of agricultural production, marketing, and pricing. The present chapter explained the application of economic theories (that we discussed in the previous chapter) in farming. It includes application of production function including Cobb-Douglas production function, input-output analysis, cost functions, etc. It also explained family-farm model and input-output pricing. A common problem faced by the farmers or producers is to decide how to allocate the resources he owns, rents, or borrows to the alternative products, which can be produced. It involves choices between the number of cows and sheep, the acres of rice and wheat, or coffee and tea, etc. A producer wishes to obtain maximum satisfaction from the use of his resources. Section on analysis of farm programmes and production possibilities unravel the theoretical underpinnings of these issues. Farmers can compare the costs of different methods of performing operations or they can compare net revenues from producing different kinds of products. In order to make these comparisons, farmers must have information in regard to the consequences of alternative choices. Construction of enterprise budgets is one method by which producer can compare directly the revenues and costs of alternative enterprises. The chapter concludes with a briefing on production of agricultural products due to changes in relative prices and costs with an explanation of the cyclical behaviour, which is demonstrated by the 'Cobweb' theorem of production.

REFERENCES

Anderson, J.R. and Hardaker, J.B. 1979. *Economics and the Design of Small-Farmer Technology*. (Ed. A.Valdes, G.M. Scobie, and J.L. Dillon) Iowa State University Press: Ames, Iowa.

Dillon, J.L. 1965. *Review of Marketing and Agricultural Economics*, 33:3, 175-189.

Dillon, J.L. 1979. *South African Journal of Agricultural Economics*. 1:1, 7-13.

Ferris, A. and Malcolm, L.R. 1999. *Proceedings of 43rd Annual Conference of the Australian Agricultural and Resource Economics Society*. Christchurch, New Zealand.

Heady, E.O. and Dillon, J.L. 1961. *Agricultural Production Functions*. Iowa State University Press: Ames.

Hochman, Z.; Coutts, J.A.; Carberry, P.S.; McCown, R.L. *Learning and Knowing Processes for Change in Agriculture in Industrialised Countries*. (Ed. B. Hubert and R. Ison) INRA: Versailles.

Morrison, D.A.; Kingwell, R.S.; Pannell, D.J. and Ewing M.A. 1986. *Agricultural Systems*, 20:4, 243-268.

Nair, Reji D. 2003. *Land Economics: Issues in Land Use*. Centre for Studies in Development Strategy. Thiruvananthapuram.

6

Farm Risk Management

Introduction

Risk is an important aspect of the farming business. The uncertainties of weather, yields, prices, government policies, global markets, and other factors can cause wide swings in farm income. Risk management involves choosing among alternatives that reduce the financial effects of such uncertainties. The present chapter discusses type of risks encounter in farming, major strategies, measures of risks and suggested programmes.

Risks in Agriculture

There are mainly five general types of risks in agriculture which are described here: production risk, price or market risk, institutional risk, human or personal risk, and financial risk :

(i) *Production risk :* It derives from the uncertain natural growth processes of crops and livestock. Weather, disease, pests, and other factors affect both the quantity and quality of commodities produced.

(ii) *Price or market risk :* Refers to uncertainty about the prices producers will receive for commodities or the prices they must pay for inputs. The nature of price risk varies significantly from commodity to commodity.

(iii) *Financial risk :* It results when the farm business borrows money and creates an obligation to repay debt. Rising interest rates, the prospect of loans being called by lenders, and restricted credit availability are also aspects of financial risk.

(iv) *Institutional risk :* This results from uncertainties surrounding government actions. Tax laws, regulations for chemical use, rules for animal waste disposal, and the level of price or income support payments are examples of government decisions that can have a major impact on the farm business.

(v) *Human or personal risk :* Refers to factors such as problems with human health or personal relationships that can affect the farm business. Accidents, illness, death, and divorce are examples of personal crises that can threaten a farm business.

Farm Risk Management Strategies in Agriculture

Farmers have many options for managing the risks they face, and most producers use a combination of strategies and tools. Some strategies deal with only one kind of risk, while others address multiple risks. Following are some of the more widely used strategies :

(i) *Enterprise diversification :* Assumes incomes from different crops and livestock activities do not move up and down in perfect correlation, so that low income from some activities would likely be offset by higher income from others.

(ii) *Financial leverage :* Refers to the use of borrowed funds to help finance the farm business. Higher levels of debt, relative to net worth, is generally considered risky. The optimal amount of leverage depends on several factors, including farm profitability, the cost of credit, tolerance for risk, and the degree of uncertainty in income.

(iii) *Vertical integration :* Generally decreases risk associated with the quantity and quality of inputs or outputs because the vertically integrated firm retains ownership or control of a commodity across two or more phases of production and/or marketing.

(iv) *Contracting* : Can reduce risk by guaranteeing prices, market outlets, or other terms of exchange in advance. Contracts that set price, quality, and amount of product to be delivered are called marketing contracts, or simply forward contracts. Contracts that prescribe production processes to be used and/or specify who provides inputs are called production contracts.

(v) *Hedging* : Uses futures or options contracts to reduce the risk of adverse price changes prior to an anticipated cash sale or purchase of a commodity.

(vi) *Liquidity* : Refers to the farmer's ability to generate cash quickly and efficiently in order to meet financial obligations. Liquidity can be enhanced by holding cash, stored commodities, or other assets that can be converted to cash on short notice without incurring a major loss.

(vii) *Crop yield insurance* : Pays indemnities to producers when yields fall below the producer's insured yield level. Coverage may be provided through private hail insurance or federally subsidized multiple peril crop insurance.

(viii) *Crop revenue insurance* : Pays indemnities to farmers based on gross revenue shortfalls instead of just yield or price shortfalls. Several subsidized revenue insurance plans are available for major crops in most areas in developed countries and also in some developing countries.

(ix) *Household off-farm employment or investment* : Can provide a more certain income stream to the farm households to supplement income from the farming operation.

Most producers use a variety of farm management strategies and tools. Risks and the willingness and ability to bear risks differ from farm to farm, so do the risk management strategies used.

Major Farm Risk Management Programmes : Required

Over the years, various government programmes have been used to achieve risk management policy goals world-wide. Some of the most important are briefed below. Unfortunately most of these programmes are not accessible to the farmers in developing countries.

(i) *Multiple peril crop insurance (MPCI) :* This is to cover yield losses from most natural causes. MPCI operated on a somewhat limited basis since the early 1980s in US and Europe, when insurance availability was greatly expanded and premium subsidies increased in hope of replacing the disaster payment programme.

(ii) *Revenue insurance :* A cousin to MPCI, it has become the most popular form of insurance in some advanced areas. Whereas crop insurance covers only yield losses, revenue insurance pays when gross revenue (yield time price) falls below a specified level.

(iii) *Disaster payments :* These are direct payments to farmers on an emergency basis when crop yields are abnormally low due to adverse growing conditions. During the 1970s, there was a "standing" disaster payments programme, with payments made without declaration of a disaster area. Regular payments ceased after 1981, but since then *ad hoc* disaster payments have been implemented in different parts of the world.

(iv) *Supplemental agricultural disaster assistance* : These are permanent disaster assistance programmes that provide payments to producers of eligible commodities (crops, livestock, farm-raised fish, and honey) for losses incurred as a result of diseases, adverse weather, or other environmental conditions. The programme for crop producers in US is called supplemental revenue assistance (SURE).

(v) *Non-insured assistance programme (NAP) :* These payments are made to producers of crops for which crop insurance is unavailable.

(vi) *Emergency loans :* These loans have been provided on various occasions to farmers as part of broad disaster assistance packages. Loans are generally repaid to the government at reduced interest rates.

(vii) *Emergency feed assistance programmes :* These have helped livestock producers in Western Europe and US to obtain feed when local pasture, hay, and forage supplies have been limited due to drought or other adverse conditions.

(viii) *Loan deficiency payments* (LDPs) : These protect producers of several major commodities against revenue losses due to low prices. LDPs pay the difference between the government's commodity loan rate and the commodity's loan repayment rate.

(ix) *Marketing assistance loans :* These allow farmers to obtain a loan for their commodity at the loan rate and repay it later at a lower loan repayment rate. The net effect is similar to collecting an LDP and selling the commodity. Most farmers prefer the LDP method over a marketing loan.

(x) *Counter-cyclical payments (CCPs) :* These are made when market prices fall below average levels. The payments may provide income risk protection to producers of several major commodities when crop prices are low.

(xi) *Average crop revenue election (ACRE):* This is an alternative to CCPs. Producers choosing to participate in the ACRE programme give up CCPs and portions of direct payments and loan programme benefits. ACRE payments are made when revenue (yield multiplied by market prices) falls below recent historical levels.

Measures of Risk

Risk refers to variability and is a complex and multi-faceted phenomenon. A variety of measures have been used to capture different facets of risk. The more important ones are *range, standard deviation, coefficient of variation, semi-variance* and *semi-standard deviation.*

To illustrate the calculation of these measures, consider an agriculture investment whose net present value has the following distribution :

NPV	*Probability*
200	0.3
600	0.5
900	0.2

The expected NPV works out to be :

$$E(NPV) = \sum_{i=1}^{3} PiNPVi$$

$$= 0.3 \times 200 + 0.5 \times 600 + 0.2 \times 900$$

$$= 540$$

Now let us discuss the various measures of risk.

(i) *Range:* Obviously the simplest measure of risk, the range of a distribution is the difference between the highest value and the lowest value. The range of the above distribution is 900 – 200 = 700.

(ii) *Standard Deviation*: Standard deviation of a distribution is :

$$\sigma = \left[\sum Pi(Xi = \overline{X})^2\right]^{1/2}$$

Where, σ = standard deviation

Pi = probability associated with the ith value

X_i = i^{th} value, and

$\overline{X}$ = expected value

The standard deviation of the NPV distribution presented above is :

$$\sigma = [0.3\,(200 - 540)^2 + 0.5\,(600\text{-}540)^2 + 0.2\,(900 - 540)^2]^{1/2}$$

$$= [62500]^{1/2}$$

$$= 250$$

The square of standard deviation is called variance;

Variance = σ^2

The variance of the above distribution is 62500.

(iii) *Coefficient of variation* : One of the problems with the standard deviation or variance is that it is not adjusted for scale. If we go only by standard deviation, an investment with an expected net present value of $ 10 million and a standard deviation of $ 100,000 would be considered more risky than an investment with an expected net present value of $ 1 million and standard deviation of $ 90,000.

The coefficient of variation (CV) adjusts standard deviation for scale. It is defined as :

$$CV = \frac{Standard\ Deviation}{Expected\ Value}$$

The coefficient of variation for the agriculture investment in our illustration is :

CV = 250/540 = 0.46

(iv) *Semi-variance :* There is yet another problem with standard deviation or variance; it considers all deviations, positive as well as negative, from the expected value in the same way. Since investors are concerned about only negative deviations, semi-variance seems to be a more suitable measure of risk. The semi-variance is computed the same way as the variance, except that only outcomes below are expected values are taken into account.

It is defined as :

$$SV = \sum Pidi'^2$$

Where d'_i is equal to d_i if $d_i < 0$ and is equal to 0 if $d_i \geq 0$

The semi variance for the investment in our illustration is :

$$SV = 0.3\,(200 - 540)^2 = 34680$$

(v) *Semi-standard deviation*: The semi-standard deviation is the square root of semi-variance. The semi-standard deviation for the investment in our illustration is :

Semi-standard deviation = (Semi Variance)$^{1/2}$

= (34680)$^{1/2}$

= 186.2

Despite the limitations the standard deviation is the most commonly used measure of risk in investment. The main reasons for using standard deviation are :

(a) If a variable is normally distributed, its mean and standard deviation contain all the information about its probability distribution;
(b) If the utility of money is represented by a quadratic function (a function commonly suggested to represent diminishing marginal utility of wealth), then the expected utility is a function of the mean and standard deviation; and
(c) Standard deviation is analytically easily tractable.

For measuring the expected value and dispersion of a variable, its probability distribution is required. In some cases the probability distribution can be defined with a fairly high degree of objectivity on the basis of past experience and evidences.

Summary

Risk is an important aspect of the farming business and there are mainly five general types of risks in agriculture: production risk, price or market risk, institutional risk, human or personal risk, and financial risk. This chapter discussed about major farm risks and the strategies required in combating such risks. Most producers use a variety of farm management strategies and tools, and since risks and the willingness and ability to bear risks differ from farm to farm, so do the risk management strategies used. Over the years, various programmes have been used by governments to achieve risk management policy goals world-wide. It includes crop and revenue insurance, disaster relief and payments, emergency loans and assistance, counter cyclical payments etc. The chapter concludes with a discussion on various measures of farm risks. The most important measures of risk include *range, standard deviation, coefficient of variation, semi-variance* and *semi-standard deviation*. Despite the limitations the standard deviation is the most commonly used measure of risk in farm investment.

REFERENCES

Dillon, J.L. 1966. *Farm Policy*, 5, 103-108.

Dillon, J.L. 1967. *The Analysis of Response in Crop and Livestock Production.* Pergamon Press.

Dillon, J.L.; Plucknett, D.; and Vallaeys, G. 1978. *Farming Systems Research at the International Agricultural Research Centres.* Unpublished report of the Technical Advisory Committee Review Team of Farming systems research at CIAT, IITA, ICRISAT and IRRI.

John M. Antle and Daniel A. Sumner. 1996. *The Economics of Agriculture: Selected Papers of D. Gale Johnson.* Vol. 1, University of Chicago Press, Chicago.

Kenneth L. Infanger. *Economics of Agricultural Management: An Introduction.* Reston Publishing Company.

McCown, R.L., 2001. *Proceedings of 10th Australian Agronomy Conference,* Hobart

Siegfried Pausewang, *et al.* 1990. *Ethiopia: Rural Development Options.* Zed Books Ltd., London.

7

Farm Business Analysis

Introduction

There is great emphasis today on record keeping on the farm. Many farmers in financial difficulties could have avoided some of these difficulties if they had records to consult. The purpose of keeping records is not just to accumulate masses of information. It is to use this information to compare and discern trends in the farm business. These trends help farmers make sensible managerial decisions. Records are useful only if they are used and simply keeping them is not sufficient.

Just as doctors examine symptoms to judge our overall health, we can look at the records of farming and assess its financial health, both now and to some extent, in the future. This financial assessment shows how the business has been doing and how we think it will do in the future. It is obviously too late to correct what has happened in the past. But we can make decisions on past trends and steer the business towards our financial objectives. This procedure is similar to a doctor telling us we have been gaining weight and suggesting a diet and exercise to steer us towards a better weight in the future.

Farm business examination will use various ratios and indicators taken from the financial statements. The most useful function of these tools is to illustrate trends over time. One year's (or one time period's) figures by themselves have limited use. A series of these figures over time shows where the business is heading and help farmers make decisions to alter course. The present chapter analyses the farm business analysis tools in detail.

Major Categories of Farm Business Analyses

There are four basic farm business analysis tools. These are *liquidity, solvency, profitability,* and *efficiency*. Each category takes most of its information from the balance sheets and income statements of the operation. The balance sheet uses two valuation columns, generally referred to as *modified cost* and *market value*. The *modified cost* column shows what we paid for our assets (including the depreciated value of any depreciable assets) and what debts we owe. The *market value* column approximates what our assets are worth in today's market (i.e., the date of the balance sheet). The *market value* of liabilities includes any income and capital gains taxes owed to the IRS if we sold the business on that date. Though we are probably not planning to sell the business, the market value of those liabilities should be included, because it shows the complete debt picture of the firm, including taxes, and will, therefore, help in intergeneration transfer plans and general tax management.

Liquidity of Farms

Liquidity is a short-term concept and shows a farm business' ability to meet debts when they become due. Many farms have liquidity problems today, meaning that they cannot pay all the principal and interest on loans due in the following year. There are three main indicators of liquidity: the *current ratio, working capital,* and the *debt structure ratio.*

Current Ratio

Liquidity indicators are found in the balance sheet. One of the most useful is the current ratio, which is a current asset over current liabilities.

$$\text{Current Ratio} = \frac{\text{Current Assets}}{\text{Current Liabilities}}$$

If, a business has a current ratio of 2 : 1, it means that there are $ 2 of current assets covering every $ 1 of current liabilities.

Agricultural lenders generally like a current ratio of at least 2 : 1. If the ratio is 1 : 1, then the farm is barely liquid, and if the ratio is less than this, the farm has liquidity problems. It is always hard to generalize in farming because of the variation between different enterprises and different areas and practices. Generally, any current ratio trends below 1.3 : 1 indicate danger. Conversely, a ratio above 5 : 1 strongly suggests that the firm is overly liquid and should be investing some of its current assets.

Working Capital

A second liquidity guide is working capital, that is, current assets minus current liabilities. Working capital shows what is available after meeting debts due.

$$Working\ Capital = Current\ Assets - Current\ Liabilities$$

Obviously, we need a positive figure; otherwise the farm is illiquid. But the amount of working capital considered reasonable depends on the size and type of the individual farm. We would expect it to grow for an expanding business. However, as a rough guide working capital should be close to expected net farm income. Otherwise, we may not be making sufficient investment in the farm.

If there is insufficient working capital, a fall in total current asset values could wipe it all out. For example, if total current assets were \$ 60,000 and working capital \$ 40,000, it would take a 67 per cent drop in asset values to eliminate the working capital. Whereas, if total current assets were \$ 150,000 and working capital still \$ 40,000, a 27 per cent fall in the value of our assets would totally deplete all our working capital.

Debt Structure Ratio

A third liquidity guide, debt structure ratio, illustrates the debt structure of the firm. This ratio is calculated by dividing current liabilities by total liabilities.

$$Debt\ Structure\ Ratio = \frac{Current\ Liabilities}{Total\ Liabilities}$$

A ratio of 0.6 : 1 (often written as 0.6) means that 60 per cent of the total farm debt is due the following year. If total debt is small, there is nothing to worry about. But most farms have considerable debt loads, and a debt structure ratio of 0.6 shows that too much of the farm debt is current. In general, a ratio of 0.2 or less is safe, and 0.5 or more is dangerous. Trends going in either direction may require some decisions. For example, we could increase borrowing in the former case and perhaps, by transferring some of the current debt into long-term liabilities, arrange for debt restructuring in the latter situation.

These liquidity ratios give some idea of the firm's current position and short-term progress. They indicate potential trouble and should be calculated regularly. But in order to examine long-term effects we need to look at solvency.

Solvency of Farms

Solvency is a long-range concept which shows the farm's ability to meet all debts when assets are sold. Solvency indicators are found in the balance sheet. The main indicators are *net worth*, the *leverage ratio*, and the *solvency ratio*.

Net Worth

The basic solvency indicator is net worth. As assets minus debt equals net worth, we are obviously looking for a positive figure. A negative net worth shows insolvency and trends in net worth show trends in solvency.

$$\textit{Net Worth} = \textit{Assets} - \textit{Debt}$$

The two-column balance sheet provides a *cost net worth* and the *market value net worth*. The *cost net worth* shows what we paid for our assets, minus any depreciation. The *market value net worth* shows what we could get for our assets if we sold them at current market prices, minus any contingent taxes. (Contingent taxes are the income and capital gains taxes which the farmer would have to pay if we sold these assets.) *Market value net worth* is generally greater than *cost net worth*. As most of the difference between the two can be attributed to inflation, we need to look at both *net worths* when examining the solvency of a business.

Leverage Ratio

The leverage ratio is another solvency indicator. This ratio is calculated by dividing total debt by net worth.

$$Leverage\ Ratio = \frac{Total\ Debt}{Net\ Worth}$$

Most lenders do not want to see leverage ratios over 1.5 : 1. This means that there is $ 1.50 of debt for every $ 1 of net worth. The higher the ratio, the more risk the farm faces, and, conversely, the lower the ratio, the lower the risk. However, many young farmers need ratios over 4 : 1 if they are to obtain sufficient capital to farm. Trends are again important. Ratios above 2 : 1 must be watched carefully. Those below 1 : 1 may suggest acquiring additional debt to exploit current opportunities. To understand the advantages and disadvantages of leverage, assume a firm is in the situation shown by the following balance sheet (Table 7.1).

Table 7.1 : Illustration of the Leverage Ratio—First Year

Assets (Rs. '000)	*December 31, 2007* *Liabilities and Net Worth (Rs. '000)*	
400	Liabilities	300
	Net Worth	100
Total 400	Total	400

The leverage ratio is 300,000/100,000 or 3 : 1, which is fairly high for a farm. Now suppose the farmer gets a 20 per cent return on his assets during the following year. This return would alter his next balance sheet as given in Table 7.2.

Farmer's 20 per cent increase in assets has resulted in an 80 per cent growth in his net worth. This is why borrowing can be so attractive. Successfully using other people's money provides impressive gains in net worth. If the farmer had no debt, he would only be controlling $ 100,000 of assets, and his assets would equal his net worth. So a 20 per cent asset return means his net worth would also grow by 20 per cent. However, by borrowing he was able to increase his return on net worth fourfold (and reduce his leverage to 1.67 : 1). This is the advantage of leverage.

Table 7.2 : Illustration of the Leverage Ratio—Second Year

Assets (Rs. '000)	*December 31, 2008* *Liabilities and Net Worth (Rs. '000)*	
480	Liabilities Net Worth	300 180
Total 480	Total	480

If in the following year provided a negative 20 per cent return on assets. The farmer's balance sheet would then read as given in Table 7.3. Even though his assets only fell by 20 per cent, his net worth has fallen to one-fifth of its original value. At the same time, his leverage ratio has increased from 3 : 1 to 15 : 1, which is an extremely dangerous level. Hence, this ratio is an important indicator of solvency, and trends in leverage should be watched.

Table 7.3 : Illustration of the Leverage Ratio—Third Year

Assets (Rs. '000)	*December 31, 2008* *Liabilities and Net Worth (Rs. '000)*	
320	Liabilities Net Worth	300 20
Total 320	Total	320

Solvency Ratio

A third guide to business solvency is the solvency ratio, found by dividing total debt by total assets.

$$Solvency = \frac{Total\ Debt}{Total\ Assets}$$

The higher the ratio, the more debt there is for each rupee of assets. Three balance sheets (Tables 7.1, 7.2 and 7.3) used to illustrate the leverage ratio, the solvency ratios were 0.75, 0.63, and 0.94, respectively. The 0.75 ratio means that there are 75 paisa of debt for every Re. 1 of assets. Any ratio approaching 1 shows imminent insolvency. Solvency, therefore, covers the long-range aspects of the business.

Profitability of Farms

We need to see how the business is performing throughout the year; which is indicated by profitability indicators taken from the income statement. Three main indicators of profitability are shown directly in the income statement. These are *net farm income, off-farm income,* and *net income*. The example given in Box 7.1 shows how they differ.

Box 7.1 : Illustration of Net Farm Income, Off-farm Income, and Net Income

- Total farm revenue (Rs. 200,000) minus total farm fixed (Rs. 40,000) and variable costs (Rs. 110,000) equal net farm income. Net farm income is, therefore, Rs. 50,000.
- Off-farm income comes from non-farm jobs and custom work on other farms, e.g., spouse earning Rs. 15,000, plus custom work paying Rs. 5,000. Off-farm income is, therefore, Rs. 20,000.
- Net income is net farm income plus off-farm income, minus all income taxes and Social Security payments due on these income sources.

Net income shows what is available to pay for the following :

(i) Principal on past debts,
(ii) New farm investments,
(iii) Family off-farm investments such as retirement funds, stocks, mutual funds, etc, (Retirement fund investing may alter the income taxes due, which were calculated earlier), and
(iv) Family living expenses.

Net income is the real bottom line of a business and is the single most important indicator of profitability. It has increased in importance in recent years as more and more money gets to the farm from off-farm sources. A model income statement is given in Table 7.4.

Table 7.4 : A Model Income Statement, January 1 to December 31, 2007

Revenue	*(Rs. '000)*
Sales	120
Inventory Change (+)	15
Other	5
Gross Revenue	140
Minus livestock for fattening and feed bought	16
Value of farm production (1)	124
Expenses	
Operating expenses	64
Depreciation	13
Interest	24
(Gain) or loss on asset sales (+)	1
Total Costs (2)	102
Income	
Net farm income (3 = 1-2)	22
Off-farm income	14
Minus income taxes and Social Security	6
Net Off-farm income (4)	8
Net Income (5 = 3+4)	30

Ratio Indicators

The remaining measures of profitability are all ratios. These are *return on assets, return on equity, growth in net worth,* and the *profit margin ratio.*

Return on Assets

Return on assets (ROA), sometimes known as the rate earned on capital investment, is calculated as follows :

$$ROA = \frac{Net\ Farm\ Income + Interest\ Paid - Family\ Withdrawal}{Beginning\ Total\ Assets}$$

Interest paid is an accrued figure. It is the amount of interest that was paid during the year, adjusted for the difference between the opening and closing accrued interest shown on the beginning and ending balance sheets. The interest paid figure is calculated on an income statement schedule and then entered on the income statement.

The figure is included in this equation because we are looking for a return on all the assets in the farm. This is because assets are generally composed of equity capital and debt capital. Assets are usually purchased with a down payment of our own money (equity capital) and borrowed money (debt capital). For example, we might buy a Rs. 60,000 tractor with Rs. 20,000 equity and Rs. 40,000 debt capital. *What does all this have to do with "interest paid": simply that we pay interest on debt capital but not on equity capital.* And interest on debt has already been subtracted as part of our costs when we calculated net farm income. We now need to add it back in order to find a total return on assets.

Family withdrawals are simply a proxy for the farmer and farm-family unpaid labour. The figure includes family living expenses and retirement account investments. Family withdrawals are calculated in the statement of owner equity, given at the end of Table 7.5 (1). The denominator is beginning total assets (Some people prefer to use an average of the opening and closing total assets as the denominator. Whichever we use, be consistent.)

Return on Equity

Return on net worth or equity (ROE) is a second useful profitability ratio. It is calculated by :

$$ROE = \frac{Net\ Farm\ Income + Family\ Withdrawals}{Beginning\ Net\ Worth}$$

This figure shows us what farmers are getting from investing their money in the farm business. If this return is less than the return they could obtain from an equal or less risky investment, then there is a good economic argument that the farmer should leave farming and invest elsewhere (It is probably just as well that most farmers use more than this economic logic in deciding where to invest.) A cost-based ROE greater than 4 per cent is particularly good.

Table 7.5 : Model balance Sheet

BALANCE SHEET (2) Dec. 31, 199__ (Rs. '000)

Assets			*Liabilities and Net Worth*		
	Cost	*Market Value*		*Cost*	*Market Value*
Current	95	100	Current	95	98
Intermediate	170	180	Intermediate	98	104
Long-term	700	980	Long-term	380	420
			Total liabilities	573	622
			Net worth	392	638
TOTAL	965	1260	TOTAL	965	1260

Statement of Owner Equity (Rs. '000)

Net worth at beginning of period (cost)	380
Net income for period	30
Gifts and inheritances	—
Additions to partnership or corporation capital	—
Total	410
Minus gifts for estate transfer	—
Minus net worth at end of period (cost)	392
Therefore, family withdrawals for period	18

Summary of Changes during Period (Rs. '000)

(1)	net worth at beginning	380
	+ net income	+30
	– family withdrawals	-18
	= net worth at the end (cost)	392
(2)	Assets decreased, or 1000-965 (cost)	35(–)
	Liabilities decreased, or 620-573 (cost)	47(+)
	Therefore, Net worth increased	12

Growth in Net Worth

Rate of *growth in net worth* is another useful profitability ratio. This ratio may initially seem more of a solvency than a profitability ratio. But net income is the deciding factor, and is our basic profitability measure. Any principal payments or investments made during the year will raise net worth. Family withdrawals will deplete it.

$$\text{Growth in Net Worth} = \frac{\text{Net Income + Family Withdrawals}}{\text{Beginning Net Worth}}$$

We obviously want a positive growth rate and, mostly, the higher the better. A good, continual and even growth in net worth shows a good business. Growth in net worth is a major indicator for a lender.

Profit Margin Ratio

The last profitability ratio we will look at is the profit margin ratio.

$$\text{Profit Margin Ratio} = \frac{\text{Net Farm Income + Interest Paid – Unpaid Labour}}{\text{Value of Farm Production}}$$

This ratio essentially shows the return to the family's management and farm capital for every rupee of farm production. The value of farm production (VFP) is gross sales plus inventory changes, minus feed purchases and purchases of livestock for fattening. VFP is simply the value added by the farm. The unpaid farm and family labour is an estimate of the salary that the people involved could earn in another available job. It is, in other words, the economic concept of opportunity cost. A figure of 11 paise shows that there is an 11 per cent return to management and capital for each Rupee one of farm production; the higher the percentage, the greater the return.

All these profitability measurements provide information on how well the business is doing. But they do not show whether the resources are used efficiently. This is where the final set of indicators comes in.

Farm Efficiency

These efficiency guides basically measure the relationship between inputs and outputs. They can be divided into *physical* and *financial measurements*. As some of the financial indices include market prices, the efficiency measurements cover the producing, marketing, and financing tasks of the business.

These are some of the most important farm management indicators.

(a) *Physical Measures*

Farmers feel familiar with most of the physical relationships of their business. The following items highlight some of the most common ones :

(i) Yield per acre;
(ii) Yield per animal in terms of births, conceptions, outputs and herd life;
(iii) Pounds of feed per pound of live-weight gain and other conversion factors; and
(iv) Machinery measurements, including miles per gallon, units per time period, performances per labour hour, and similar efficiency factors.

These figures come from the basic physical records of the business.

(b) *Financial Measures*

The financial measures of efficiency are all ratios. These include the *operation ratios*, the *debt service ratio*, and the *times interest earned ratio*.

The Operation Ratios

The operation ratios are a set of five ratios, all using the same denominator, namely, the value of farm production, VFP. The ratios come from the income statement. Summing the numerators gives the VFP, so all the ratios collectively add to 1 or to 100 per cent. They are as follows :

$$(1)\ \text{The Operations Ratio} = \frac{\text{Total Expenses} - \text{Depreciation} - \text{Interest}}{\text{VFP}}$$

$$(2)\ \text{The Operations Ratio} = \frac{\text{Depreciation}}{\text{VFP}}$$

$$\text{(3) The Operations Ratio} = \frac{\text{Interest}}{\text{VFP}}$$

$$\text{(4) The Operations Ratio} = \frac{\text{Net Farm Income}}{\text{VFP}}$$

$$\text{(5) The Operations Ratio} = \frac{\text{Gain/Loss on Asset Sales}}{\text{VFP}}$$

The fifth ratio is usually the residual. If there is a gain in selling farm assets, then this ratio must be subtracted from the sum of the previous four ratios in order to get 100 per cent. Likewise, if there is a loss in sales, the ratio is added to the other four. In the former case, the gain has already been counted in the net farm income, so it must now be subtracted to avoid double counting. In the latter case, the depreciation reflects a higher book figure for those assets sold at a loss, so we add losses back to increase returns by decreasing costs.

For example, if the depreciated value of a machine is Rs. 1,000 and we sold it for Rs. 1,300, then we have a gain of Rs. 300. Conversely, if the sale only brought Rs. 600, then we have lost Rs. 400. Recent trends have shown that interest represents an increasing proportion of costs. Operating expenses will probably represent the lion's share of these ratio proportions. So farmers usually try to increase output to counter any cost increase.

Debt Service Ratio

The next two ratios are concerned with debt. The debt service ratio shows the proportion of the VFP that pays for annual principal and interest payments on intermediate and long-term assets.

$$\text{Debt Service Ratio} = \frac{\text{Principal + Interest on Capital Assets}}{\text{VFP}}$$

The recent trend has been for this proportion to grow as farmers attempt to service their past debt fulled expansions with lower commodity prices. This ratio is one of the most important for agricultural lenders. Many of them would be reluctant to lend anything but operating capital (i.e., for current assets) to farms where the proportion is over 33 per cent.

Times Interest Earned Ratio

The final ratio we will look at is the *times interest earned ratio.* There are several definitions of this ratio; however, the one used here is probably the most useful for farmers. It shows how much money is available to pay interest.

$$\textit{Times Interest Earned Ratio} = \frac{\textit{Net Income} + \textit{Interest Paid} - \textit{Family Withdrawals}}{\textit{Interest Due}}$$

Recall that net income is the bottom line of the income statement. It shows what is available for family living expenses, for new investment, and for principal payments. Assuming that family withdrawals must continue, regardless of the firm's performance, we add back the interest paid to show the whole amount available to pay interest. Interest paid is the same as interest due if the manager meets all his interest bills.

If the ratio is 2 : 1, this means that there is Rs. 2 available for every Re. 1 of interest due. If the ratio is less than 1, the farmer has insufficient money available to pay his interest. Hence, this ratio is sometimes called the *embarrassment ratio.*

An example of farm analysis is given in the Appendix to this chapter.

Cost of Production Budgeting (COP Budgeting)

Choosing what crops or livestock to produce is an essential decision of any farm business. One critical factor in making that decision is the cost of producing the "enterprises" being considered. This is known as enterprise budgeting or cost of production budgeting. Enterprises are a single crop or livestock commodity that produces a marketable product. Cost of Production (COP) budgeting consists of estimating the costs associated with an enterprise and the expected revenue. This section outlines the process and use of COP budgeting for farm-level decision-making.

While the format of COP budgets can vary they typically include the following components :

(i) *Revenue :* The gross revenue from crop or livestock sales before any expenses have been deducted.
(ii) *Direct Variable Costs* : Expenses for the production of a specific commodity. These change depending on the level of production (i.e., seed, fertilizer, pesticides and feed).
(iii) *Indirect Variable Costs* : Expenses used in producing all commodities on the farm (i.e., fuel, labour and utilities). These also change depending on the level of production.
(iv) *Fixed Costs :* Expenses that remain the same regardless of the level of production (i.e., property taxes, fire insurance and depreciation).
(v) *Net Profit (loss)* : revenue minus all variable and fixed costs.

Cost of Production (COP)

Knowing cost of production is vital for making farm level decisions. Following are the important points to remember while reviewing the cost of production :

(i) *Enterprise mix decisions :* It is advisable to review cost of production for each individual farm for the past three to five years. This shows how much each is contributing to the whole farm financial picture, illustrating which enterprise is making money and which is not.
(ii) *Purchasing and marketing decisions :* Pricing targets for inputs and outputs can be set at different cost breakeven levels. Knowing the breakeven points allows taking advantage of buying or selling opportunities when they arise. Use the following formulas to determine breakeven points.

$$\textit{Breakeven price to cover variable costs} = \frac{\textit{Total variable costs}}{\textit{Expected yield}}$$

= Rs./unit produced

This is the minimum price needed to cover variable costs

$$\textit{Break even price to cover total costs} = \frac{\textit{Total costs}}{\textit{Expected yield}}$$

= Rs./unit produced

This is the minimum price needed to cover all costs.

$$\textit{Breakeven yield} = \frac{\textit{Total costs}}{\textit{Expected price}} = \textit{Unit produced}$$

This is the minimum yield needed to cover all costs.

(iii) *Investment decisions :* Making the right investments in capital assets like land, machinery and buildings is critical to long-term success. COP information shows what the farm can afford to pay for those assets. Review investments in enterprises that fail to meet total costs in the long run, and redirect resources to more profitable enterprises.

Preparation of COP Budget

The cost of production information starts with good farm records. If current record-keeping system does allow estimating costs at farm level are to modified to suit the modern estimation methods. Farm accounting software makes it easier to track this level of information. The following points are very important in preparing COP budgets.

(i) *Historical Estimates* versus *Projections* : Budgeting estimates use either historical estimates or projections. Historical estimates use actual farm results and provide a measure of past enterprise performance. Projections are estimates for a future period.

(ii) *Unit Produced* versus *Production Unit :* There are two units of measurement to base costs in preparing a budget—unit produced or production unit. *Unit produced* is the output or yield that the commodity is commonly measured by, for example kilograms of corn, market hogs sold or tonnes of grapes. Production unit is the base unit used to produce the output or yield such as a hectare of land, square foot of greenhouse space, sow, cow or ewe. Since the production unit remains fairly constant throughout production, budgets are usually prepared on a production unit basis. First calculate the Rupees dollar per production unit; then test this with different yield levels to calculate Rupees per unit-produced costs.

(iii) *Estimating Farm Revenue* : Historical estimates use the average price received multiplied by the average yield to arrive at farm revenue. Projections can use average historical price/yield or expected price/yield, depending on the purpose of the budget. To obtain a good planning estimate, use: the past five-year average for prices and yields, or a season-adjusted market price for the month of the expected time of sale.

Box 7.2 : Illustration of Estimating Farm Revenue

Average grape yield per hectare × Average market price
= Grape Farm revenue
17.5 tonnes per hectare × Rs. 555.00 per tonne = Rs. 9,712.50 per hectare

(iv) *Estimating Farm Costs* : The difficulty many farmers have in COP budgeting is allocating costs to the specific farms. And the more farms there are, the more difficult the allocation process. First identify farms and include them all, even those used by other enterprises, such as hay fed to livestock (Calculate the home-grown feed enterprise costs to determine if it is cheaper to grow feed on the farm or to purchase it from off-farm sources). Assign costs to the individual enterprises. Then, use the following main approaches to estimating enterprise costs : *using on farm records, market value information* and *formula based.*

Using On-farm Records

The on-farm records approach can be further broken down into *direct costing* and *allocating whole farm expenses. Direct costing* uses the actual amount paid for each cost item by farms. This requires commodity-specific records. Direct costing is the preferred cost estimation method for direct variable costs because it uses actual prices and quantities, providing the best estimate of farm's COP. *Allocating whole-farm expenses* distributes each item to a commodity according to an allocation scheme. It is difficult to allocate whole farm expenses as they are not directly attributable to a single farm enterprise, but all enterprises share the cost. Whole farm expenses include items such as advertising, utilities, memberships/subscription fees, office expenses, professional fees, motor vehicle expenses, small tools, property taxes, and fire and liability insurance premiums. We must allocate whole farm expenses to arrive at a total cost of production for a farm enterprise. But this should have no effect on enterprise selection decisions. Farm choices should be the same after allocating whole farm expenses as it was before they were added.

Common allocation methods of whole farm expenses involve the following components :

(a) *Per cent of enterprise gross margin* : This is the recommended method. For example, if a commodity accounts for 30 per cent of the whole farm gross margin, the commodity is charged 30 per cent of the overhead costs. Use one of the other methods below if there are objective reasons to do so.

(b) *Per cent of sales :* Expenses are allocated according to how much the enterprise contributes to the gross sales.

(c) *Per cent of total expenses* : Expenses are allocated according to how much the enterprise contributes to the whole farm expenses.

(d) *Number of hectare (acre)-trips :* That is no-till field operations require less machinery passes so the machinery cost is less than conventional tillage.

(e) *Hours spent* : Allocating labour costs across the enterprises. If machinery hours are tracked by enterprise this can be effective way to allocate these costs.
(f) *Self defined* : Based on experience and unique to site.

No one allocation method is right for every farm business or every enterprise or for all expenses; it may take a variety of methods. Use the method that makes the most sense for farm and each expense. Once allocations are developed and, provided there are no significant changes to the enterprise mix, keep them consistent. This will make for better year-to-year comparisons.

Market Value

It is used for costs such as land, labour and machinery. The preferred estimate for land costs is the cash rental rate, since cash rental rates are assumed to cover all costs associated with owning agricultural land. Labour requirements on the farm can be supplied by a combination of paid and unpaid labour. To estimate costs for all labour multiply the total number of hours needed (paid and unpaid) in an enterprise by the current hourly wage rate. Machinery costs can be difficult to estimate on a farm enterprise level. A common approach is to use the custom farm-work rates for the operations involved in the farm enterprise to arrive at the enterprise machinery cost. The downside to using market values is they may not accurately reflect what is happening on the farm. There can be large differences between market prices and own costs. Use the records to provide the most accurate measure of costs.

Formula Based

A formula-based approach is particularly useful for estimating capital costs associated with farm machinery and buildings. This method takes fuel use and repair rates, replacement costs and years of expected life to insert into formulas that calculate annual variable and fixed costs.

There is nothing wrong with using different methods for different individual costs. Table 7.6 presents Estimation Approaches by Enterprise Expense on an individual cost basis.

Table 7.6 : Estimation Approaches by Enterprise Expenses—Individual Cost Basis

Direct	*Allocate Whole Farm Expenses*	*Market Value*	*Formula-based*
1. Commodity purchases (seed, livestock, feed grain)	1. Motor vehicle expenses	1. Labour	1. Machinery (gasoline, diesel, fuel, oil)
2. Fertilizers and soil supplements	2. Small tools	2. Agricultural Contract work	2. Machinery (repairs, licences, insurance)
3. Pesticides and chemical treatments	3. Containers and twine	3. Freight and shipping	3. Building and fence repairs
4. Prepared feed, minerals and salts	4. Soil testing	4. Commissions and levies	4. Depreciation (buildings and machinery)
5. Custom feeding	5. Office expenses/legal and accounting fees	5. Storage/drying	5. Interest (operating)
6. Veterinary fees, medicine, breeding fees	6. Advertising and and pro-motion costs	6. Rent (land, buildings, pastures)	6. Interest (real estate, mortgage, term loans)
7. Insurance premiums (crop or production)	7. Memberships/ subscriptions/ licences/ permits	7. Machinery lease/rental	
8. Other crop and livestock supplies	8. Motor vehicle interest and leasing costs		
	9. Utilities (electricity, telephone, heating fuel) Market Value		
	10. Other insurance premiums		

Other Cost Considerations

(a) *Establishment costs* : Perennial crops only produce after an "establishment" period. The costs associated with this non-productive time need to be recovered during the crops productive life. It includes an estimate of the annual cost to recover establishment costs in the annual budget of the years in full production. Spread these costs over the expected productive life of the crop.

(b) *Depreciation* : It is non-cash cost that measures the loss of value of machinery or buildings over time. It is the portion of the cost of the machinery or building that is counted as an expense each year. Costs are spread over their expected useful life. The depreciation rate used for tax purposes typically does not reflect the expected useful life of the asset. A rate that better reflects the life of the asset is more appropriate for business management purposes.

(c) *Interest : Interest paid on loans* is the actual cash paid in interest for existing loans. This gives a good estimate for cash flow purposes but fails to account for the cost of having money invested in the operation. *Interest on investment* represents the cost of money tied up in purchasing assets or operating inputs whether the money is borrowed or farmer's own. This is also referred to as the opportunity cost; it is what the farmer could have earned with that money if it had invested in the next best alternative. The interest rate used should reflect conservative rates of return for money that could be obtained in the current market. This assigns an interest cost to farm investments regardless of whether the money is borrowed or not.

(d) *Property taxes :* Land costs are a good example. If there is no land debt there are no cash interest costs; but there is a cost associated with having money invested in land and not available for other uses. Use the formula-based method to calculate interest on investment.

Interpreting COP Budget

The COP budget measures the following three different margins :

Contribution Margin

Contribution margin is revenue minus direct variable costs or production related costs. For crops this includes seed, fertilizer, pesticides, crop insurance and marketing. For livestock this includes feed, livestock purchases, custom feeding, health, breeding and marketing. This is the simplest margin to calculate since we are dealing with costs directly related to growing crops or raising livestock. If the contribution margin is negative at this point, think seriously about whether this will ever be a profitable enterprise. Costs would need to be significantly reduced or revenue significantly increased to make it worthwhile—and there are still many costs that have not yet been included. Use contribution margin as a quick reference point in making short term cropping decisions. If a new crop does not require additional labour or machinery, the direct variable costs of the crop choices will change the most from one crop to the next. The contribution margin tells us whether the enterprise is contributing to the other indirect costs and by how much compared to other enterprises.

Gross Margin

Gross margin is revenue minus all variable costs. This involves allocating whole farm costs that all enterprises must share. Use the gross margin to decide if it makes sense to invest in capital assets for this enterprise. Capital assets are long-term assets with a useful life longer than a year. These include land, buildings, machinery and quota.

If the gross margin is negative, determine if anything can be realistically done on the cost or revenue side to turn this into a positive. A positive number indicates that the enterprise does contribute to paying some or all of its fixed farm costs. If, for example, the enterprise covers all variable costs and only a portion of fixed costs, we could continue with the enterprise in the short-run, while looking at either cost reductions or revenue increases to have it cover all costs in the future.

> **Box 7.3 :** Illustration of Investment in Land
>
> Interest on investment costs = Land value x Interest on investment rate
> Land value—Land is valued at its current fair-market value for agricultural purposes (e.g. Rs. 1,400/hectare).
> Interest on investment rate—Use conservative, current market investment rates (e.g. 3.5%)
> Interest on investment on land = Land value x Interest rate = Rs. 1,400 x 3.5% = Rs. 49.00 per hectare

Profit Margin

Profit margin is revenue minus all variable and fixed costs. Without long-term profit, a farm business is not sustainable. Sustainability depends on every enterprise covering all costs and providing a return to management.

Risk and the COP Budget

Farming is a risky business and profit is a return to risk (Refer previous chapter). To make a profit, the farmer must take risk as we discussed in previous chapter. The goal is to learn to manage it. It is necessary to develop reasonable expected yield, cost and price estimates in decision making. However, expected outcomes are not enough for effective decision making. A producer must also consider the consequences of outcomes other than those expected or are considered most likely. Since a wide variety of risks can affect yield, price and cost, it is highly unlikely that actual net returns will match our estimated net returns. Understanding and anticipating these unpredictable risks, is critical to budgeting.

Cost of production information is an essential ingredient for farm level decision making. Knowing the cost of production is the first step in controlling them. Good cost of production information starts with good farm records. If the current record keeping system does not lend itself to estimating costs on an enterprise level, start making changes at the earliest.

Summary

The farm business analysis shows how the business has been doing and how it will do in the future. It is obviously too late to correct what has happened in the past. But we can make decisions on past trends and steer the business towards our financial objectives. Farm business examination will use various ratios and indicators taken from the financial statements. There are four basic farm business analysis categories: *liquidity, solvency, profitability,* and *efficiency.* Each category takes most of its information from the balance sheets and income statements of the operation. In this chapter we discussed about three liquidity indicators, three solvency indicators, four profitability indicators and two efficiency indicators in major categories. Choosing what crops or livestock to produce is an essential decision of any farm business. One critical factor in making that decision is the cost of producing the "enterprises" being considered. This is known as enterprise budgeting or cost of production budgeting, which consists of estimating the costs associated with an enterprise and the expected revenue. This Fact sheet outlines the process and use of COP budgeting for farm-level decision-making. This chapter also covers preparation and interpretation of cost of production budget. This chapter concludes with a brief account of risk and its impact on the COP budgeting. Since a wide variety of risks can affect yield, price and cost, it is highly unlikely that actual net returns will match our estimated net returns. Understanding and anticipating these unpredictable risks, is critical to budgeting.

REFERENCES

Anderson, J.R. 1974. *Review of Marketing and Agricultural Economics,* 42:1, 3-55.

Bakken, B.; Gould, J. and Kim, D. 1994. *Modelling for Learning Organisations.* (Ed. J.D.W. Morecroft and J.D. Sterman) Productivity Press: Portland, Oregon. pp. 243-266.

Frey, Thomas L. 1980 "Financial Statements: Developing a Framework for Analysis," *Agri Finance.* Skokie, IL.

Kenneth L. Infanger. *Economics of Agricultural Management: An Introduction.* Reston Publishing Company.

Malcolm, L.R. 1990. *Review of Marketing and Agricultural Economics,* 58:1, 24-52.

McCown, R.L.; Carberry, P.S.; Foale, M.A.; Hochman, Z.; Coutts, J.A.; Dalgliesh, N.P. 1998. *Proceedings of 10th Australian Agronomy Conference,* Wagga Wagga, pp. 633-636.

Stephen B. Harsh, *et al., Managing the Farm Business.*

Van Blokland, P.J. *Introducing the Balance Sheet,* Extension Circular No. 651. Florida Cooperative Extension Service, Institute of Food and Agricultural Sciences, University of Florida, Gainesville, FL 32611.

Van Blokland, P.J. *Introducing the Income Statement,* Extension Circular No. 645. Florida Cooperative Extension Service, Institute of Food and Agricultural Sciences, University of Florida, Gainesville, FL 32611.

Van Blokland, P.J. *Introducing the Cash Flow,* Extension Circular No. 656. Florida Cooperative Extension Service, Institute of Food and Agricultural Sciences, University of Florida, Gainesville, FL 32611.

Van Blokland, P.J. *Introducing the Statement of Change in Financial Position,* Extension Circular No. 658. Florida Cooperative Extension Service, Institute of Food and Agricultural Sciences, University of Florida, Gainesville, FL 32611.

Van Blokland, P.J. *Linking the Financial Statements,* Extension Circular No. 657. Florida Cooperative Extension Service, Institute of Food and Agricultural Sciences, University of Florida, Gainesville, FL 32611.

APPENDIX

Illustration of Farm Analysis

Example

The following consists of seven tables: Balance Sheet—1; Income Statement; Liquidity; Solvency; Profitability; Efficiency; and a combined summary table of Balance Sheet—2, Statement of Owner Equity, and Summary of Changes during Period.

Major Developments during the Period

The Rs. 12,000 shown in the summary below is the difference between the net income and family withdrawals for the year (i.e., Rs. 30,000–Rs. 18,000). It is equivalent to the business term "residual earnings" and shows what was available during the year for buying assets and reducing liabilities. This Rs. 12,000 is also the increase in net worth, as shown earlier. Stated another way, net worth at the beginning of the period plus residual earnings equals the end of period net worth.

So the business generated Rs. 12,000. But liabilities fell by Rs. 47,000. How was this done? It was largely fulled by the Rs. 35,000 decrease in assets, of which Rs. 5,000 came from current and the remaining Rs. 30,000 from intermediate assets. The current decrease probably stemmed from selling inventory, while the intermediate reduction was most likely due to depreciation. Depreciation adds cash by reducing taxes. This is why an after tax cash flow is more valuable than a pre-tax cash flow.

Analysis

Start by examining the broad picture of the business, as shown by its assets, liabilities, net worth, and income. *What happened in each category? Why did it happen? What do we do about it?*

Table 7.A.1 : Balance Sheet (1) Dec. 31, 19__ (Rs. '000)

Assets			*Liabilities and Net Worth*		
	Cost	*Market Value*		*Cost*	*Market Value*
Current	100	110	Current	100	105
Intermediate	200	210	Intermediate	120	130
Long-term	700	1000	Long-term	400	440
			Total liabilities	620	675
			Net worth	380	645
TOTAL	1000	1320	Total	1000	1320

Table 7.A.2 : Income Statement Jan. 1—Dec. 31, 19__ (Rs. '000)

Revenue	*Rs.*	
Sales	120	
Inventory Change (+)	15	
Other	5	
Gross Revenue	140	
Minus livestock for fattening and feed bought	16	
Value of Farm Production	124	(1)
Expenses		
Operating expenses	64	
Depreciation	13	
Interest	24	
(Gain) or loss on asset sales (+)	1	
Total Costs	102	(2)
Income		
Net farm income (1–2)	22	(3)
Off-farm income	14	
Minus income taxes and Social Security	6	
	8	(4)
Net Income (3+4)	30	

Firstly look at liquidity, solvency, profitability, and efficiency. See the liquidity ratios and particularly the trends in each ratio. *What happened? Which way is the trend going? What action should we take?*

Table 7.A.3 : Liquidity

		Cost (Rs. '000)	*Market Value (Rs. '000)*
Current Ratio	= current assets / current liabilities	100/100 = 1	110/105 = 1.05
Working Capital	= current assets–current liabilities	100–100=0	110–105 = 5
Debt Structure	= current liabilities/ total liabilities	100/620 = 0.16	105/675 = 0.15

Follow the same procedure with solvency and then combine the liquidity and solvency findings into a complete picture of the structure of the business. *What does this picture tell us? Is the trend in structure meeting our objectives for this business? What do we need to do?*

Table 7.A.4 : Solvency

	Cost (Rs. '000)	*Market Value (Rs. '000)*
Net Worth = assets – debt	1000 – 620 = 380	1320 – 675 = 645
Leverage Ratio = total liabilities/ net worth	620/380 = 1.63	675/645 = 1.04
Solvency Ratio = total liabilities/ total assets	620/1000 = 0.62	675/1320 = 0.68

Table 7.A.5 : Profitability

NFI (net farm income) Rs. 22,000	OFI (off-farm income) = Rs. 14,000		NI (net income) = Rs. 30,000	
Ratio Indicators				
	Cost (Rs. '000)	*(%)*	*Market Value (Rs. '000)*	*(%)*
ROA (return on assets) = (NFI + interest paid – family withdrawals)/ beginning total assets	(22 + 24 – 18)/ 1000	= 2.8	(22 + 24 – 18)/ 1320	= 2.1
ROE (return on equity) = (NFI – family withdrawals)/ beginning net worth	(22 – 18)/380	= 1.1	(22 – 18)/645	= 0.6
Growth in Net Worth = (NI – family withdrawals)/ beginning net worth	(30 – 18)/380	= 3.2	(30 – 18)/645	= 1.9
Profit Margin = (NFI + interest paid – unpaid labour*)/value of farm production (VFP)	(22 + 24 – 15)/ 124	= 25		

* Estimated at Rs. 15,000

The next step is to see what this structure is producing in terms of profitability and efficiency. Run through each ratio and see what the trends show. It may be possible to find comparative figures for some of the efficiency ratios. Many of them are production oriented, and land-grant institutions tend to specialize in this area. Discussing your findings with other people will certainly help in acquiring expertise in this new and unfamiliar arena called farm business analysis.

Table 7.A.6: Efficiency

Operation Ratios			
		(Rs. '000)	*(%)*
1.	(total expenses – depreciation – interest)/VFP	(102 – 13 – 24)/124	= 52
2.	depreciation/VFP	13/124	= 10
3.	interest/VFP	24/124	= 18
4.	NFI/VFP	22/124	= 19
5.	(gain) or loss on asset sales/VFP	1/124 = 1	
Debt Service Ratio	(principal + interest on capital assets*) / VFP	(20 + 22 + 12)/124	= 0.43
Times Interest Earned Ratio	(NI + interest paid – family withdrawals)/interest due	(30 + 24 – 18)/24	= 1.5

* interest estimated at Rs. 12,000; annual principal payments are Rs. 20,000 for long-term debts and Rs. 22,000 for intermediate debts.

Table 7.A.7: Balance Sheet (2) December 31, 199__ (Rs. '000)

Assets			*Liabilities and Net Worth*		
	Cost	*Market Value*		*Cost*	*Market Value*
Current	95	100	Current	95	98
Intermediate	170	180	Intermediate	98	104
Long-term	700	980	Long-term	380	420
			Total liabilities	573	622
			Net worth	392	638
Total	965	1260	Total	965	1260

(contd.)

(Table: 7.A.7 *contd.*)

Statement of Owner Equity (Rs. '000)

Net worth at beginning of period (cost) .		380
Net income for period		30
Gifts and inheritances		—
Additions to partnership or corporation capital		—
	Total	410
Minus gifts for estate transfer		—
Minus net worth at the end of period (cost)		392
Therefore, family withdrawals for period		18
Summary of Changes during Period (Rs. '000) :		
1 Net worth at beginning		380
+ Net income		+30
– Family withdrawals		–18
= Net worth at the end (cost)		392
2 Assets decreased, or 1000-965 (cost)		35(–)
Liabilities decreased, or 620-573 (cost)		47 (+)
Therefore,Net worth increased		12

Problem

The present exercise is meant to show a typical farm, which has liquidity problems but is solvent. Its net worth is growing, although this growth relies heavily on off-farm income. It appears to be managed efficiently and can pay its bills, despite a high annual interest payment. *What advice you would give if you are a farm business manager?*

Table 7.B.1: December 31, 1984

Assets (Rs. '000)		*Liabilities and Net Worth (Rs. '000)*	
	400	Liabilities	300
		Net Worth	100
Total	400	Total	400

Table 7.B.2 : December 31, 1985

Assets (Rs. '000)		*Liabilities and Net Worth (Rs. '000)*	
	480	Liabilities	300
		Net Worth	180
Total	480	Total	480

Table 7.B.3 : December 31, 1985

Assets (Rs. '000)		*Liabilities and Net Worth (Rs. '000)*	
	320	Liabilities	300
		Net Worth	20
Total	320	Total	320

Table 7.B.4

Net Farm Income	50,000
Off-Farm Income	+ 20,000
Income Taxes	– 10,000
Social Security	2,800
Net Income	= 57,200

Look at the broad categories first and consider what you would do about liquidity.

8

Agriculture Planning and Project Analysis

Introduction

Economic Planning means deliberate attempt on the part of a country or government to follow a specific pattern of economic growth and structural change in the economy. Planning generally implies intervention by government in the normal working of the market. However, the degree of intervention very much depends on government ideology. For instance, France follows a system of indicative planning whereby the degree of intervention is very minimal. Indeed, government simply sets some targets for the private sector, outlines the major directions of economic policy, and provides information to the market economy so that the desired targets can be achieved. In the Socialist system the government controls most of the means of production, dictates the pattern of demand and supply, and owns the 'commanding' heights of the economy. As a consequence, the degree of state intervention in the market mechanism is the greatest. In between these two extremes we have a large number of intermediate degrees of government intervention in the operation of the market.

Planning Techniques

Input-Output Analysis

The use of input-output (IO) analysis has now become quite common as a sectoral planning tool in many countries. The technique was originally developed by Leontiff, and is based on the input-output tables showing the linkages between various sectors (agriculture, industry, etc.) of the economy and providing the foundation of the economy-wide general equilibrium model that focuses on production. The following are the important assumptions of IO analysis :

(a) Inputs are needed in constant proportion to produce output in each sector;
(b) Each activity or sector can produce only one product; and
(c) Joint products cannot be produced by any sector.

The first assumption implies that the returns to scale are constant. It is necessary to separate prices and quantities if we wish to develop a model from the IO tables. Assuming a closed economy, that is, no exports and imports, the rows of IO tables can be written as follow :

$$P_i X_i = \Sigma P_i X_{ij} + P_i F_i \qquad \text{.......... (1)}$$

Where X_i production in sector i, P_i price of output in sector i, X_{ij} flow of intermediate goods from sector i to j, F_i final demand for sector i. Let a_{ij} intermediate needs from sector i per unit of output of sector j.

As the IO coefficient is fixed we have

$$a_{ij} = X_{ij}/X_j \qquad \text{.......... (2)}$$

Hence, in any period, we have

$$\frac{P_i a_{ij}}{p_j} = \frac{P_i x_{ij}}{p_j x_j} \qquad \text{.......... (3)}$$

With an IO table for a given year, it is useful to state the units of real flows in such a way that prices will add up to one.

If we divide the equation (1) by the price, we have

$$X_i = \Sigma a_{ij} X_j + F_i \qquad \text{.......... (4)}$$

In matrix form this material balance equation is :

$$X = AX + F \qquad \text{.......... (5)}$$

And hence we can solve for X as follows :

$$X = (1 - A)^{-1} F \qquad \text{.......... (6)}$$

Note that the solution indicates that given the final demand, production for different sectors can be determined so that supply is equal to demand in different sectors.

This model can be extended to include foreign trade. Assume that *E* is exports and *M* imports. Then we have

$$X + M = A\,X + F + E \qquad \text{.......... (7)}$$

Which gives

$$X = (1 - A)^{-1} (F + E - M) \qquad \text{.......... (8)}$$

Notice that IO technique is very useful if planners wish to promote a balanced growth between agriculture and industry. Notice also that many LDCs are net importers of food and as such the use of equation-8 raises problem as the net final demand turns to be negative.

One of the major limitations of the IO technique is that it is not helpful in finding the optimal technique of production from a variety of technologies that may be available for producing a given commodity. In order to identify the optimal technology, it is necessary to use linear programming models. It is also to note that the IO analysis is a macro-planning tool, and its relevance to sectoral planning for agriculture is largely confined to testing for the feasibility and consistency of sectoral targets.

Linear Programming

It is a mathematical technique for the analysis of optimum decisions, subject to certain constraints in the form of linear inequalities. Linear programming is defined as a method to decide the optimum combination of factors to produce a given output or the optimum combination of products to be produced by given plan and equipment. Mathematically speaking, it applies to those problems which require the solution of maximization or minimization problems subject to a system of linear inequalities stated in terms of certain variables. The problems of maximization and minimization are also called optimization problems. It agriculture planning it is mainly used to decide between a variety of techniques and equipment.

In farm planning, the farmers have to decide whether to use labour-intensive or capital intensive technique of production, depending on outlay. They will choose that technique which maximizes output. Let us assume that it is planned to produce a farm commodity *Z* using *X* and *Y* inputs, its objective is to maximize output. It has two alternative production processes, *C* (capital intensive) and *L* (labour intensive). The constraint is a given cost outlay *MP* as shown in Figure 8.1.

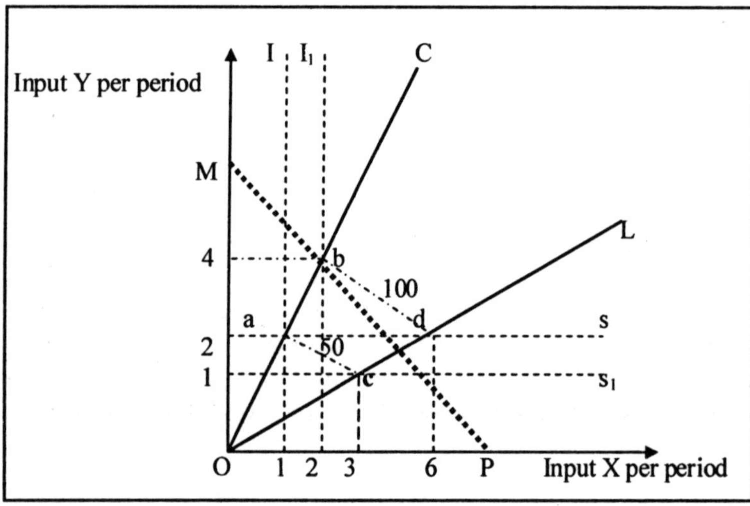

Fig. 8.1 : Feasible Solutions under Linear Programming

If process *C* requires 2 units of input *Y* to every unit of input *X*, it will produce 50 units of commodity *Z*. If the inputs *X* and *Y* are doubled to 4 units of *Y* and 2 units of *X*, output is also doubled to 100 units of *Z*. These combinations of *X* and *Y*, represented by *a* and *b*, establish the output scale along the capital intensive process ray OC. On the other hand the labour intensive ray OL is represented by the input combinations *c* and *d*. When the points *a* and *c* at the 50 units output level on the linear rays OC and OL are joined, they form an isoquant $Iacs_1$; at the 100 units output level the corresponding isoqunt is I_1bds. The cost outlay constraint is represented by the isocost line MP and it places a limit on the production capacity of the project. The project can produce with either of the two available techniques C and L within the area represented by the triangle Obd. It is not possible for it to produce outside the *area of feasible solutions*. The *optimal solution* which maximizes the output will occur at the point where the isocost curve touches the isoquants I_1bds at point *b* on the process ray OC. It shows that the project will use the capital intensive technique C by using 4 units of input Y and 2 units of input X and produce 100 units of output Z.

If we take another farm enterprise project whose objective functions is to maximize its revenue subject to certain constraints of limited capacities. Assume it produces two products, X and Y and has four departments such as manufacturing, assembling, polishing and packing of the product, each with a fixed capacity and are designated as A, B, C and D. This problem is illustrated graphically in Figure 8.2.

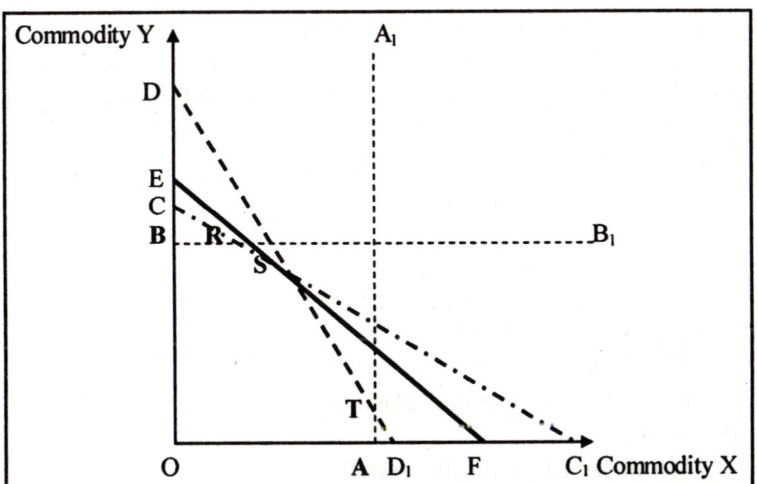

Fig. 8.2 : Feasible Production under Linear Programming

The production of X and Y is subject to constraints A, B, C and D. Constraint A limits the production of X on OA, constraint B limits the production of Y on OB, constraint C limits the production of both X and Y to OC_1 and OC, and constraint D limits their production to OD_1 and OD respectively. The area OATSRB shows all combinations X and Y that can be produced without violating any constraints. This is the area of *feasible production* within which X and Y can be produced, but there is no possibility of producing any combination at any point outside this area.

The original solution can be found out by taking an isoprofit line within the feasibility zone. An isoprofit line represents all combinations of X and Y which yield the same profit to the firm. The optimal solution lies on the highest isoprofit line EF on the polygon OATSRB. This is the point S and all other points lie outside the zone of *feasible production.*

Every linear programming maximization problem has its dual problem, that of minimization. The original problem is known as the primal problem, which always has its dual. If the primal problem pertains to maximization, the dual involves minimization, and *vice versa.*

Suppose the farmers undertake a project which aims at minimization of costs. Two types of goods X_1 and X_2 are to be produced. Let the planners attach weights of 3 and 8 to units of these goods. Let the production of 2 units of X_1 use 1 unit of input C_1 and 2 units of input C_2; similarly for the production of X_2 we use 2 units of C_1 and 8 units of C_2. This problem can now be set in the linear programme for as:

Maximize $3X_1 + 8X_2$ (*R* – revenue)

Subject to constraints:

$X_1 + 2X_2 \leq 2$ and

$2X_2 + 8X_2 \leq 6$ and none of these quantities is negative. The optimal solution is $X_1 = 2$, $X_2 = ½$ and $R = 7$.

The dual problem is:

Let P_1 be the imputed price of X_1 and P_2 be the imputed price of X_2 :

Minimize $2P_1 + 6P_2$ (C—cost)

Subject to the constraints:

$P_1 + 2P_2 \geq 3$ and

$2P_1 + 8P_2 \geq 8$ and that none of these prices is negative. The optimal solution is:

$P_1 = 1,\ P_2 = ½$ and $C = 7$.

These are the dual prices, but as all values have been imputed to the two resources, the maximum value of the objective function *C* must equal *R*.

Hence, $C = R = 7$.

As given in Figure 8.3, line *AB* represents $P_1 + 2P_2 = 3$ and the line *CD* represents $2P_1 + 8P_2 = 8$. The feasible solution lies on or above the line AZD. The optimal solution is at point Z where the isocost *RK* passes through the point of intersection of *AB* and *CD*.

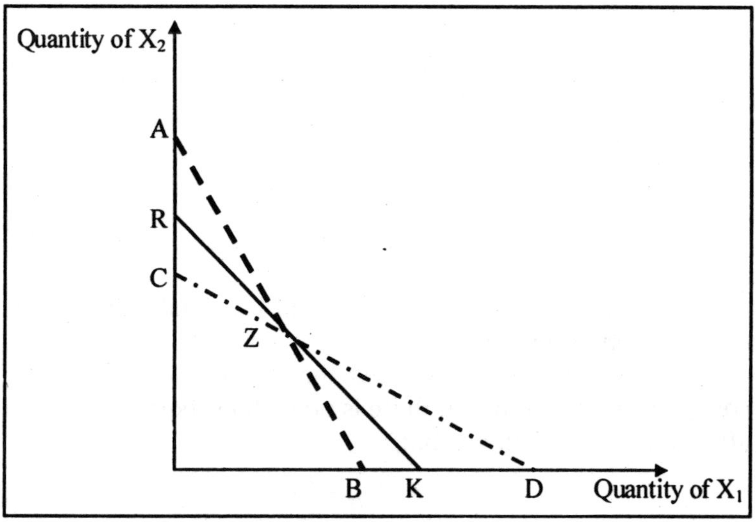

Fig. 8.3 : Optimal Solutions under Linear Programming

Agriculture Project Planning

Project appraisal is a form of applied welfare economics or cost-benefit analysis. Its central objective is to estimate the social net benefits of an investment project on the economy as a whole in contradiction to merely estimating its financial prospects as viewed by private investors. The earliest examples of cost-benefit analysis were concerned with the appraisal of major flood control, irrigation and hydro-electric schemes and it was realized that there are many hidden costs and benefits associated with such schemes.

Two of the most comprehensive decision criteria used in project appraisal are the net present value (NPV), and the internal rate of return (IRR). A third useful criterion, the benefit-cost ratio (*B/C*), is similar to *NPV*.

Net Present Value (NPV)

NPV is the difference between the PV of total benefits and the PV of total costs. That is;

$$PV_{TB} = \sum_{t=1}^{T} \frac{B}{(1+r)^t} \qquad \text{.......... (9)}$$

Where *B* is total benefits. Similarly,

$$PV_{TC} = K_0 + \sum_{t=1}^{T} \frac{V}{(1+r)^t} \qquad \text{.......... (10)}$$

Where K_0 is initial investment and *V* is total annual costs of operation and maintenance.

For simplicity, we assume that both *B* and *V* remain constant for all *t*, and that there is no further fixed investment after time *t*=0. It follows that;

$$NPV = PV_{TB} - PV_{TC} \qquad \text{.......... (11)}$$

$$= \sum_{t=1}^{T} \frac{B}{(1+r)^t} - \left[K_0 + \sum_{t=1}^{T} \frac{V}{(1+r)^t} \right] \qquad \text{.......... (12)}$$

Equation 12 expresses the difference between the *PVs* of two time streams, the one representing benefits, the other costs.

In use, the condition *NPV*>0 must be satisfied to justify the project. Competitive project are ranked by the magnitude of *NPV*, subject to the previous condition unless there is an overall capital constraint. The discount rate is exogenous, that is it is part of the input data.

Internal Rate of Return (IRR)

IRR is the discount rate satisfying the condition $NPV = K_0$, where $K_0 \geq 0$. Thus we have

$$K_0 + \sum_{t=1}^{T} \frac{B - V}{(1+r)^t} \qquad \text{.......... (13)}$$

Where *i* is IRR. In use the expected IRR must reach some positive minimum value to justify the project. More formally, justification requires satisfaction of the condition IRR$\geq i^t$, where i^t is the opportunity cost of capital, the social time preference rate, or other qualifying discount rate. Competitive projects are ranked by the magnitude of IRR, subject to the previous condition. The discount rate *i* is endogenous, that is it emerges from the analysis. However, the test discount rate, i^t, still has to be determined independently.

Benefit-Cost Ratio (B/C Ratio)

The B/C ratio calculation is identical with those of the NPV criterion. But instead of taking $PV_{TB} - PV_{TC}$ we take PV_{TB}/PV_{TC}. More specifically :

$$B/C = \sum_{t=1}^{T} \frac{B}{(1+r)^t} \left[K_0 + \sum_{t=1}^{T} \frac{V}{(1+r)^t} \right]^{-1} \qquad \text{.......... (14)}$$

In use a project is justified provided B/C $\geq$1. Competitive projects are ranked by the magnitude of B/C subject to the first condition. Unlike the NPV criterion, this one does not discriminate between large and small projects.

One of the most critical problems of project appraisal is the choice of discount rate. The test discount rate is implicit in all NPV and B/C ratio estimates. It also defines the minimum acceptable rate or return for comparison with IRR estimates. Pragmatically, if the discount rate is set too high a bias is created in favour of projects with relatively low initial capital expenditure and with benefits accruing earlier rather than later. So in agriculture, for example, a rapidly maturing extension project might be favoured at the expense of an irrigation project with a larger initial capital requirement and a longer lead time. If the discount rate is set too low the converse arguments applies. Additionally, the choice of a low rate will tend to result in too many projects being justified.

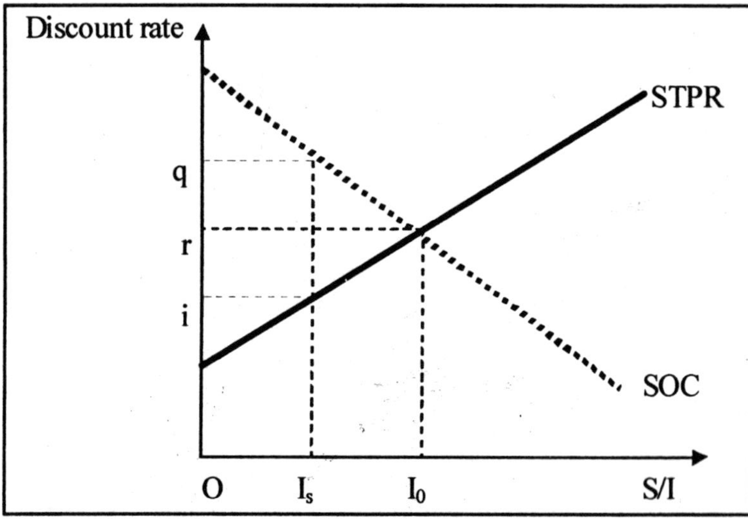

Fig. 8.4 : Choice of second-best discount rate

Theoretically the optimum discount rate is the one which equates the marginal social opportunity cost of capital (SOC) with the marginal social time preference rate (STPR). Whereas SOC defines the rate of trade-off between present and future consumption, according to the level of current saving and investment, STPR shows the rate at which society collectively trades off present against future consumption. Since it is economically rational to rank investment projects in descending order of their expected rate of return, an inverse relationship between the levels of saving/investment and SOC is assumed. The reward for waiting goes down with the declining social profitability of ranked projects. The relationship between the levels of S/I and STPR is assumed to be direct, reflecting the increased marginal opportunity cost of waiting at higher levels of S/I. The two functions are shown in Figure 8.4, where given a socially sub-optimum level of investment I_s, the SOC is q whereas the STPR is lower at i. The condition SOC = STPR defines the optimum level of investment I_0 and the discount rate r. Thus, where the coordinates of SOC and STPR meet at I_0 and r, the gap between q and i is effectively eliminated.

In the real world a government may be unable to induce as much investment as I_0 due to a saving gap. The government is more farsighted than private investors and taxpayers, particularly regarding investment for the benefit of future generations. Although private citizens as taxpayers may be prepared to sacrifice more, that is accept a higher cost and lower reward for waiting than as voluntary savers, the total level of saving and investment is nevertheless subject to political constraints. Thus if the first-best level of saving and investment and the corresponding social equilibrium discount rate are I_0 and r, it is easy to see from Figure 8.4 that a compromise, synthetic discount rate may be a second best solution. So, if I_s is the highest feasible level of total saving and investment, q and i, respectively, represent the upper and lower limits of the synthetic rate d (that is, $q>d>1$). Regarding d as some weighted average of q and i, it can be argued that the relative weights depend upon national economic priorities and political judgement. Being fairly closely related to the opportunity cost of capital in the private sector, q is likely to be easier to estimate by observation than i (which is much more subjective). Thus in practice, choosing d may boil down to little more than answering the question: 'How much less is d than q?'

Summary

Agriculture planning has become an important tool in both developed and developing countries. The most common planning tools like input-output analysis and linear programming are quite common in farm planning. The use of input-output (IO) analysis has now become quite common as a sectoral planning tool in many countries. One of the major limitations of the IO technique is that it is not helpful in finding the optimal technique of production from a variety of technologies that may be available for producing a given commodity. In order to identify the optimal technology, it is necessary to use linear programming models. This chapter also discussed about the important project appraisal tools which can be used in agriculture. Project appraisal is a form of applied welfare economics or cost-benefit analysis. Its central objective is to estimate the social net benefits of an investment project on the economy as a whole in contradiction to merely estimating its financial prospects as viewed by private investors. The most comprehensive decision criteria used in project appraisal are the net present value (NPV), the internal rate of return (IRR), and the benefit-cost ratio (B/C).

REFERENCES

Anderson, J.R., Dillon, J.L., Hardaker, J.B. 1985. *Agricultural Systems Research for Developing Countries.* (Ed. J.V. Remenyi). Australian Centre for International Agricultural Research (ACIAR): Canberra. pp. 77-88.

Dillon, J.L. 1976. *Agricultural Systems.* 1, 5-22.

Evan Drummond, H. and John W. Goodwin. 2001. *Agricultural Economics*, Prentice Hall, New Jersey.

Harold G. Halcrow. 1984. *Agricultural Policy Analysis.* McGraw Hill Book Company. New York.

John M. Antle and Daniel A. Sumner. 1996. *The Economics of Agriculture : Selected Papers of D. Gale Johnson.* Vol. 1, University of Chicago Press, Chicago.

Martin Upton. 1996. *The Economics of Tropical Farming Systems.* Cambridge University Press.

Petheram, R.J. and Clark, R.A., 1998. *Aust. J. Exp. Agric.*, 38, 101-115.

9

Supply Response in Agriculture

Introduction

As we discussed earlier farming is one of the sectors which is mostly open to frequent risks. It ranges from climatic fluctuation to price variations and the structural transformation of economies. The present chapter discusses supply response of agriculture from the perspective of price.

The Cobweb Model

Supply response in underdeveloped agriculture generally means the variation of agricultural output and acreage mainly due to the variation in price. Let Q_a be the volume of agricultural output and P indicates the price level, W_t weather condition (*e.g.* rainfall), A acreage and t any time period. A typical simple supply response function can be written as:

$$Q_a = f(P_t - 1, A_t, W_t, U_t, \text{..........}) \qquad \text{..........} (1)$$

Where, P_{t-1} is really a proxy for the expected price and U_t a statistical error term. Such a supply response function indicates that the volume of agricultural output will depend upon past price level of such output, the acreage under cultivation, at present and the level of rainfall in the current year plus a catch-all variable which is not specifically known, that is Y_t—the yield variation—is implicit in U_t. The exact nature of such a function can vary due to difference in the nature of the economies. But at the outset, the reason behind formulating such a function is quite clear.

It is to be expected that output will vary positively with past price and land under cultivation but it could either rise or fall with changes in rainfall depending upon whether or not we have a normal rainfall or flood and drought.

Sometimes, acreage under cultivation is taken as a proxy for output due to a very close and direct relationship between acreage and output. That is,

$$Q_{at} = a_1 + a_2 A_t \qquad a_2 > 0 \qquad \text{..........} (2)$$

Hence in many cases, supply response has been assumed to be equivalent to response of acreage under cultivation to changes in economic and non-economic factors. Thus we have,

$$A_t = f(P_{t-1}, W_t, \text{..........}, U_t) \qquad \text{..........} (3)$$

Sometimes, equation-3 has been modified to take into account the impact of relative rather than absolute prices. Assume that P_{at} stands for the price of alternative crop that can be raised from the same plot of land, whereas P_{0t} is the own price of the crop. Hence we obtain :

$$A_t = f(P_{0t-1}/P_{at-1}, W_t, \text{.........}, U_t) \qquad \text{..........} (4)$$

The mechanism of the supply response can be illustrated with the help of the Figure 9.1, where we measure acreage (A) on the horizontal axis and price on the vertical axis. The curve OA represents the supply response of acreage to changes in level of prices. Such a curve is clearly upward sloping. In the bottom section of the figure we show the relationship between output and acreage. As it has been already suggested, such a relationship is positive and this can be approximated by the curve OC. It can be easily shown from the figure that when the price is OP_1, the amount of acreage under cultivation is OA_1 and output is OQ_1. If the price level rises to OP_2, acreage under cultivation rises to OA_2 and output increases to OQ_2.

However, it is possible to postulate a direct relationship between price of agricultural output and its supply. However, we can write:

$$Q_{at} = f(P_{at-1}) \qquad \text{..........} (5)$$

Or

$$Q_{at} = f\,(P_{0t-1}/P_{at-1}) \qquad \text{.......... (6)}$$

Such a supply response model is generally known as the Cobweb model.

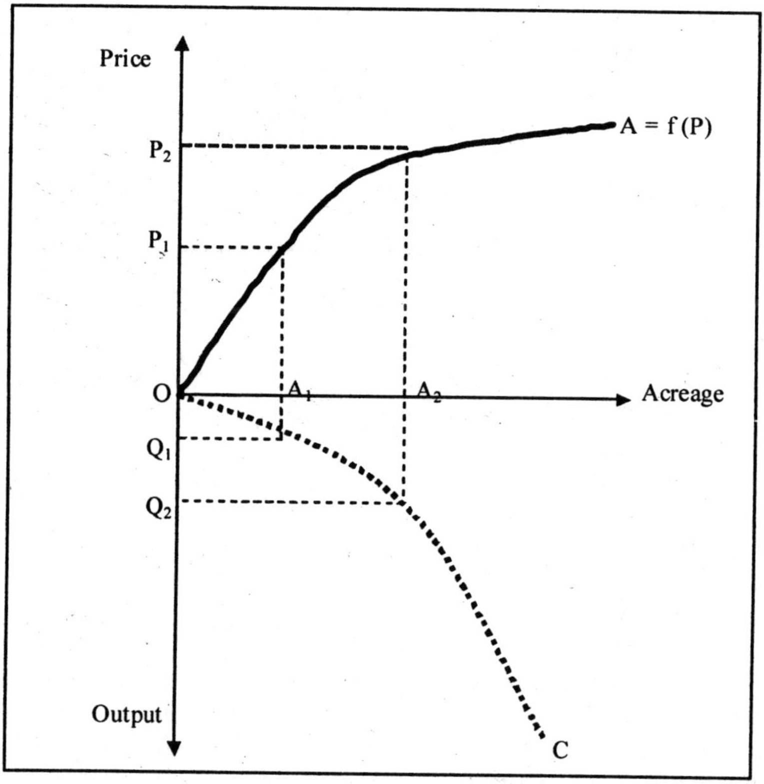

Fig. 9.1 : Output and Acreage Response to Price

Cobweb Model—Illustration

The Cobweb model is based on the theory of lagged adjustment. Given the annual cycle of agricultural crops, markets in agriculture operate period by period and as such it is cyclical. If this is true then the continuous adjustment principle should be replaced by a discrete one in the following way :

$$P_t - P_{t-1} = f\,[D_p\,(P_{t-1}) - S\,(P_{t-1})\,] \qquad \text{.......... (7)}$$

If the value of '*f*' were found to be very high, then the price amplitude of fluctuations, after an initial disturbance, could be ever-increasing around the equilibrium price and this could render the model 'explosive' (that is, once the equilibrium price has been disturbed, there will be a tendency to move away from the point of equilibrium). Let us define equilibrium as the point where demand (D_t) is equal to supply (S_t). Hence we have,

$$S_t = D_t \qquad \text{.......... (8)}$$

S_t is not instantaneous supply curve but lagged output function. Let the demand function be written as,

$$D_t = a_0 - a_1\,P_t \qquad \text{.......... (9)}$$

This simply means that demand is inversely related to price, and as price goes up, demand goes down and *vice versa.* The form of the demand curve is assumed to be linear to facilitate exposition.

The supply curve can be written as a function of the price in the last period (P_{t-1}) since it is assumed that current supply decisions are taken solely on the basis of past prices. Thus we have on the basis of our rather naïve assumption,

$$S_t = b_0 + b_2\,P_{t-1} \qquad \text{.......... (10)}$$

Such a model is recursive as it is argued that current year's supply is given by previous year's price and such a supply determines current price, given the current demand and market-clearing conditions. Once we obtain current year's price, we can determine supply in the next year. Such a situation is shown in Figure 9.2.

It is obvious from the diagram that the equilibrium price will be given at P^e price and output will be OQ_0. Let us assume that a drought leads to a crop failure and we have OQ_1 amount of output. Given an excess demand the price level rises to OP_1. Such an increase in price level encourages farmers to produce OQ_2. But then we have an excess supply, price level falls to OP_2 which, in its turn, leads to a reduction of supply to OQ_3.

At this output we have an excess demand situation which results in a rise in price level. The oscillation process ends when the equilibrium price and output are obtained at OP^e and OQ_0.

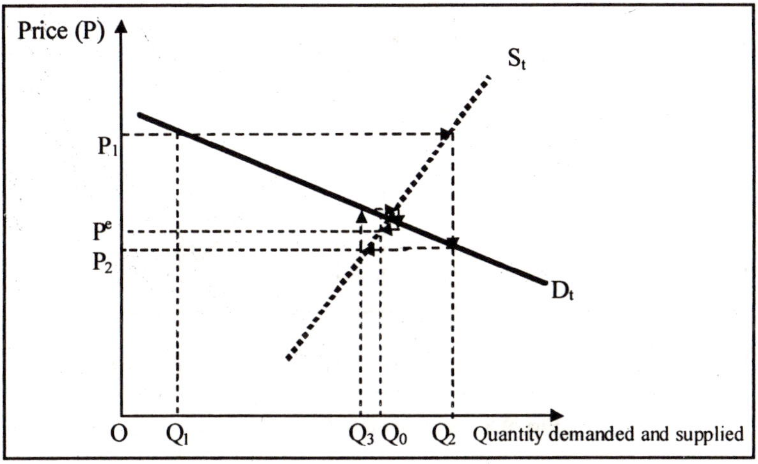

Fig. 9.2: A Convergent Cobweb Model

It is necessary to point out that both price and output can overshoot (above the equilibrium values) or undershoot (below the equilibrium values) due to inherent mechanism of a Cobweb theorem. Notice that in Figure 9.2 the amplitude of fluctuations in price and output gradually falls and eventually convergence (that is movement towards the equilibrium) occurs at the equilibrium. This is due to the elasticities of demand and supply curves. There is no reason to assume that a Cobweb cycle will always converge. Indeed in can diverge, in which case there will be a movement away from the equilibrium values. This is demonstrated in Figure 9.3 simply by changing the slopes of the *D* and *S* curves.

The Cobweb cycle may also be continuous, which implies no movement towards equilibrium or away from the initial cycle (See Figure 9.4). If the Cobweb movements are divergent, then the market is unstable. If it is convergent, then the market is stable.

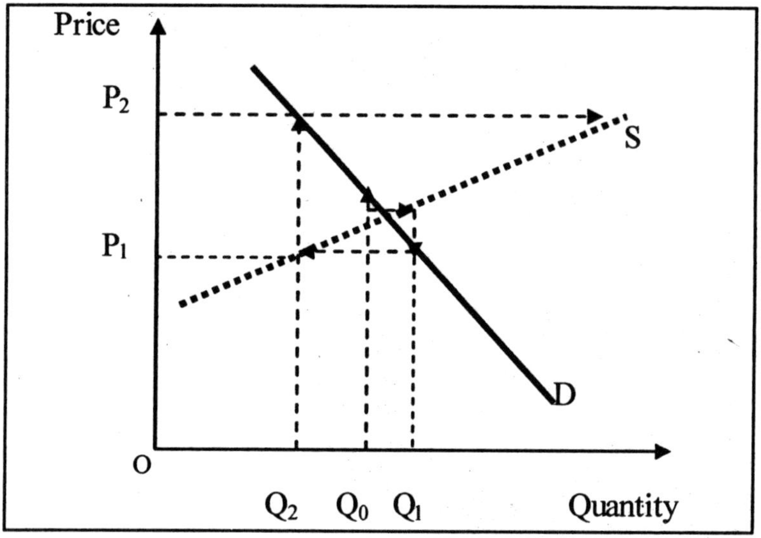

Fig. 9.3 : An Unstable Cobweb Model

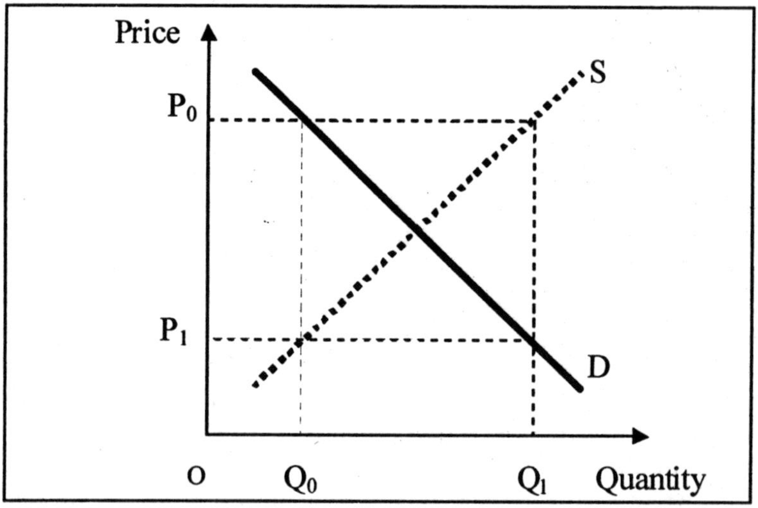

Fig. 9.4 : A Neutral Cobweb Model

A Simple Supply Response Model

It is possible to develop a simple supply response model to analyze the impact of price on output. Such a study usually involves the specification of output equation with reference to only two variables. Let us assume that total output (*Q*) of any crop is given by the product of acreage (*A*) and yield (*Y*). Thus we have,

$$Q = A.Y \qquad \text{.......... (11)}$$

Let us assume that both acreage under cultivation and yield respond to price (*P*) changes. Total differentiating we get,

$$\partial Q/\partial P = Y.\ \partial A/\partial P + A.\ \partial Y/\partial P \qquad \text{.......... (12)}$$

Assuming constant returns to scale (that is, an increase in input will raise output by the same proportion) and dividing the above equation by *Q/P*, we get,

$$\frac{dQ/dP}{Q/P} = \frac{Q/A.\partial A/\partial P}{Q/P} + \frac{Q/Y.\partial Y/\partial P}{Q/P} \qquad \text{.......... (13)}$$

$$\text{Or } \ell qp = \frac{\partial A/\partial P}{A/P} + \frac{\partial Y/\partial P}{Y/P} \qquad \text{.......... (14)}$$

$$\text{Or } \ell qp = \ell ap + \ell yp \qquad \text{.......... (15)}$$

Where ℓap is the elasticity of acreage with respect to price, ℓyp the elasticity of yield with respect to price, and ℓqp the elasticity of output with respect to price.

It is generally assumed that ℓyp is non-negative. To obtain a lower bound estimate of ℓqp it is only necessary to regress acreage on price. In empirical studies some have assumed that the price elasticity of yield is negligible. Nerlove (1958) have emphasized the possibility of important yield responses to prices in countries like the USA, where substantial increases in yields have substituted for a limit on acreage.

The idea of regressing acreage rather than output on price is not very difficult to understand. Acreage is really a proxy variable since planned output (the variable to be explained) cannot be directly observed. However, it is possible to use realized output as a proxy for planned output. But realized output could differ significantly from planned output due to the impact of random factors (e.g. Weather) on agricultural output. However, acreage is generally under the direct control of the cultivator and hence it is usually accepted as the proxy variable for output.

As regards yield response and technical change, it should be mentioned that the substitution of yield for crop area in the West has been substantially due to adoption of higher yielding crop varieties (as well as to higher fertilizer application) and the same could be said of some LDCs during the 1960s and 1970s due to Green Revolution.

But, in the short run, without technical change, yield response is likely to be confined to variation in the intensity of harvesting, etc. For example, if the price at the harvest is very low (that is, below the marginal cost of harvesting) farmers may not bother to harvest at all. But such varieties in short-term yield response may not be picked up empirically by using only annual data.

Role of Expected Price in Supply Response

It is important to bear in mind that acreage planted could vary with expected price of the output. Hence, it may be necessary to develop a model which includes expected price as one of the explanatory variables. The price explanation models have been developed by Cagan (1956) and Nerlove (1958), *inter alia,* in the economic literature. Following Cagan, elasticity of expectation (ℓex) can be defined as the percentage change in expected future price divided by percentage change in present prices.

$$\ell ex = \frac{P_{t+1} - P_t}{P_t} / \frac{P_t - P_{t-1}}{P_{t-1}} \qquad \text{..........} (16)$$

Hence a unit elasticity of expectation will imply that the proportion of expected price change to present price change will be the same. However, in Cagan's model, expectations are adaptive, which means people change their expectations in proportion to the error related with previous levels of expectation. Expected price (P_t*) will thus be the sum of past expected price (P^*_{t-1}) plus a proportion (β) of the difference between actual past price (P_{t-1}) and the past expected price. Hence we can write;

$$P^*_t = P^*_{t-1} + \beta [P_{t-1} - P^*_{t-1}] \qquad \text{.......... (17)}$$

$$0 < \beta \leq 1$$

Where, β is coefficient of price expectation associated with price uncertainty. It can be shown that,

$$P^* t = \sum_{t=0}^{T} \beta(1-\beta)^t P_{t-1-i} \text{..........(18)} \qquad \text{.......... (18)}$$

Since, $P^*_t = (1 - \beta) P^*_{t-1} + \beta P_{t-1}$ .... (19)

$= \beta P_{t-1} + (1 - \beta) [(1 - \beta)P^*_{t-2} + \beta P_{t-2}]$ .... (20)

and so on.

The above equation shows that anticipated prices form a geometrically falling lag structure as a function of all past prices (but cannot be estimated empirically because we are unable to observe P^*_{t-1}, P^*_{t-2}, etc).

When output is a function of anticipated prices, we have,

$$Q_t = a + bP^*_t + u_t \qquad \text{.......... (21)}$$

It is clear from the last three equations that we have,

$$Q_t = a + b\, \Sigma\beta\, (1-\beta)^i\, P_{t-1-i} + u_t \qquad \text{.......... (22)}$$

To estimate the above equation, it is necessary to follow an interactive procedure to iterate β in order to obtain the highest proportion of explained variation (by the independent variables) to total variation.

In Nerlove's model, anticipated prices determine equilibrium output (Q^*_t) and in each production period, output is partially altered in proportion to the difference between last periods actual output and the long-run equilibrium output. Hence, we have the following basic equations :

$$Q^*_t = a + bP^*_t \qquad \text{.......... (23)}$$

$$P^*_t = \beta P_{t-1} + (1 - \beta) P^*_{t-1} \qquad \text{.......... (24)}$$

$$Q_t = \gamma Q^*_t + (1 - \gamma) Q_{t-1} \qquad \text{.......... (25)}$$

Where γ is the rate of adjustment associated with technical and institutional rigidity.

Manipulation to eliminate non-observable variables gives,

$$Q_t = \beta\gamma a + \beta\gamma b P_{t-1} + [(1-\beta) + (1-\gamma)]\, Q_{t-1} - (1-\beta)(1-\gamma)\, Q_{t-2} \quad \text{..........} (26)$$

Equation 26 can be re-written as;

$$Q_t = d + e\, P_{t-1} + f\, Q_{t-1} - g\, Q_{t-2} \quad \text{..........} (27)$$

Where e and $e/1\text{-}f$ are the short-run and long-run supply elasticity parameters. Since β and γ cannot be separately identified, equation 27 can be solved empirically only by assuming either that $\beta = 1$ or that $\gamma = 1$ (which is obviously restrictive). This assumption enables the lag 2 taken in Q_t to be ignored. The Nerlove's supply response model is obviously more general than the simple supply response model.

Summary

Supply response in underdeveloped agriculture generally means the variation of agricultural output and acreage mainly due to a variation in price. The Cobweb model is based on the theory of lagged adjustment. Given the annual cycle of agricultural crops, markets in agriculture operate period by period and as such it is cyclical. The Cobweb cycle may be continuous, which implies no movement towards equilibrium or away from the initial cycle; if the Cobweb movements are divergent, then the market is unstable and if it is convergent, the market is stable. This chapter also illustrates a simple supply response model as well. The acreage planted could vary with expected price of the output. Hence, the chapter concluded with a model which includes expected price as one of the explanatory variables.

REFERENCES

Abbott, J.P Makeham. 1990. *Agricultural Economics and Marketing in the Tropics*. Longman Group.

Bishop, C.C. and W.D. Toussaint. 1958. *Introduction to Agricultural Economic Analysis*. John Wiley & Sons. New York.

Dillon, J.L. 1967. *The Analysis of Response in Crop and Livestock Production.* Pergamon Press.

Evan Drummond, H. and John W. Goodwin. 2001. *Agricultural Economics.* Prentice Hall, New Jersey.

John M. Antle and Daniel A. Sumner. 1996. *The Economics of Agriculture: Selected Papers of D. Gale Johnson.* Vol. 1, University of Chicago Press, Chicago.

Laurens van der Laan, H. *et al.* (eds). 1999. *Agricultural Marketing in Tropical Africa.* African Studies Center. Leiden.

Peter G. Helmberger and Jean Paul Chavas. 1996. *The Economics of Agricultural Prices.* Prentice Hall, New Jersey.

Peter G. Helmberger and Jean Paul Chavas. 1996. *The Economics of Agricultural Prices.* Prentice Hall, New Jersey.

Bibliography

Abbott, J.P. Makeham. 1990. *Agricultural Economics and Marketing in the Tropics*, Longman Group.

Anderson, J.R. 1974. *Review of Marketing and Agricultural Economics*, 42:1, 3-55.

Anderson, J.R., Dillon, J.L., Hardaker, J.B. 1985. *Agricultural Systems Research for Developing Countries*. (Ed. J.V. Remenyi) Australian Centre for International Agricultural Research (ACIAR): Canberra. pp. 77-88.

Anderson, J.R. and Hardaker, J.B. 1979. *Economics and the Design of Small Farmer Technology*. (Ed. A.Valdes, G.M. Scobie, and J.L. Dillon) Iowa State University Press: Ames, Iowa. pp. 11-26.

Bakken, B.; Gould, J. and Kim, D. 1994. *Modeling for Learning Organisations*. (Ed. J.D.W. Morecroft and J.D. Sterman) Productivity Press: Portland, Oregon. pp. 243-266.

Bennett, D. and Syme, G.J. 1979. *A Report on the Computer Farm Management Planning Service of the Farm Management Service Laboratory of Western Australia* (Inc.) CSIRO Division of Land Resources: Perth.

Bishop, C.C. and W.D. Toussaint. 1958. *Introduction to Agricultural Economic Analysis*. John Wiley & Sons. New York.

Case, H.C.M. and Williams, D.B. 1957. *Fifty Years of Farm Management*. University of Illinois Press: Urbana.

Collinson, M. 2000. *A History of Farming Systems Research*. (Ed. M. Collinson) CAB International: Wallingford, UK. pp. 5-11.

Dent, J.B. and Anderson, J.R. 1971. *Systems Analysis in Agricultural Management*. (Ed. J.B. Dent and J.R. Anderson) Wiley: Sydney. pp. 3-14.

Dillon, J.L. 1965. *Review of Marketing and Agricultural Economics*, 33 : 3, 175-189.

Dillon, J.L. 1966. *Farm Policy*, 5, 103-108.

Dillon, J.L. 1967. *The Analysis of Response in Crop and Livestock Production*. Pergamon Press.

Dillon, J.L. 1976. *Agricultural Systems*. 1, 5-22.

Dillon, J.L. 1979. *South African Journal of Agricultural Economics*. 1:1, 7-13.

Dillon, J.L.; Plucknett, D. and Vallaeys, G. 1978. *Farming Systems Research at the International Agricultural Research Centres*. Unpublished report of the Technical Advisory Committee Review Team of Farming Systems Research at CIAT, IITA, ICRISAT and IRRI.

Doyle, C.J. 1990. *Systems Theory Applied to Agriculture and the Food Chain* (Ed. J.G.W. Jones and P.R. Street) Elsevier Applied Science: London. pp. 89-112.

Evan, Drummond H. and John W. Goodwin. 2001. *Agricultural Economics*. Prentice Hall. New Jersey.

Ferris, A. and Malcolm, L.R. 1999. *Proceedings of 43rd Annual Conference of the Australian Agricultural and Resource Economics Society*. Christchurch. New Zealand.

Frey, Thomas L. 1980. "Financial Statements: Developing a Framework for Analysis." *Agri Finance*. Skokie, IL.

Harold G. Halcrow. 1984. *Agricultural Policy Analysis*. McGraw Hill Book Company. New York.

Heady, E.O. and Dillon, J.L. 1961. *Agricultural Production Functions*. Iowa State University Press: Ames.

Hochman, Z.; Coutts, J.A.; Carberry, P.S.; and McCown, R.L. *Learning and Knowing Processes for Change in Agriculture in Industrialised Countries*. (Ed. B. Hubert and R. Ison) INRA: Versailles.

John, B. Penson, Oral Capps and C. Parr Rosson. 2002. *Introduction to Agricultural Economics*. Prentice Hall: New Jersey.

John, M. Antle and Daniel A. Sumner. 1996. *The Economics of Agriculture: Selected Papers of D. Gale Johnson*. Vol. I, University of Chicago Press. Chicago.

Johnson, G.L. 1981. *International Agricultural Programs : Small Farms in a Changing World : Prospects for the 1980s*. Kansas State University: Manhattan.

Kenneth, L. Infanger. *Economic of Agricultural Management : An Introduction*. Reston Publishing Company.

Kingwell, R.S. and Pannell, D.J. (Eds.) 1987. MIDAS, *A Bioeconomic Model of a Dryland Farm System.* Pudoc: Wageningen.

Laurens van der Laan, H., *et al.* (eds). 1999. *Agricultural Marketing in Tropical Africa.* African Studies Center. Leiden.

Ladd, Hollist W., and F. Lamond Tullis (eds). 1987. *Pursuing Food Security : Strategies and Obstacles in Africa, Asia, Latin America and the Middle East.* Lynne Rienner Publishers. London.

Makeham, J.P. and Malcolm, L.R. 1993. *The Farming Game Now.* Cambridge University Press: Cambridge. Melbourne.

Malcolm, L.R, 1990. *Review of Marketing and Agricultural Economics,* 58 : 1, 24-52.

Malcolm, L.R. 2000. *Proceedings of 44th Annual Conference of the Australian Agricultural and Resource Economics Society.* Sydney.

Martin, Upton. 1973. *Farm Management in Africa.* Oxford University Press.

Martin, Upton. 1996. *The Economics of Tropical Farming Systems.* Cambridge University Press.

McCown, R.L., 2001. *Proceedings of 10th Australian Agronomy Conference.* Hobart

McCown, R.L.; Carberry, P.S.; Foale, M.A.; Hochman, Z.; Coutts, J.A.; and Dalgliesh, N.P. 1998. *Proceedings of 10th Australian Agronomy Conference.* Wagga Wagga. pp. 633-636.

Michael D. Bonehilise and Veronon R. Edanan. 1984. *Farm Management.* John Wiley & Sons.

Morrison, D.A.; Kingwell, R.S.; Pannell, D.J. and Ewing M.A. 1986. *Agricultural Systems.* 20 : 4, 243-268.

Musgrave, W.F. 1976. *Australian Journal of Agricultural Economics.* 20 : 3. 133-144.

Norman, D. 2000. *A History of Farming Systems Research.* (Ed. M. Collinson) CAB International : Wallingford, UUUK. pp. 293-299.

Pannell, D.J. 1996. *Review of Agricultural Economics.* 18, 373-383.

Pannell, D.J.; Malcolm, L.R. and Kingwell, R.S. 2000. *Agricultural Economics.* 23 : 1, 69-78.

Peter, G. Helmberger and Jean Paul Chavas. 1996. *The Economics of Agricultural Prices*. Prentice Hall. New Jersey.

Petheram, R.J. and Clark, R.A., 1998. *Aust. J. Exp. Agric.* 38, 101-115.

Ronald, D. Kay and William M. Edwards. 1999. *Farm Management*. McGraw-Hill Company.

Reji D. Nair. 2003. *Land Economics: Issues in Land Use*. Centre for Studies in Development Strategy. Thiruvananthapuram.

Shaner, W.W.; Philipp, P.F.; and Schmehl, W.R. *Farming Systems Research and Development*. Guidelines for Developing Countries. Westview Press. Colorado.

Siegfried, Pausewang, *et al.* 1990. *Ethiopia: Rural Development Options*. Zed Books Ltd. London.

Stephen, B. Harsh, *et al.*, *Managing the Farm Business*.

Taylor, J. 1994. *Fashioning Farmers: Ideology, Agricultural Knowledge and the Manitoba Farm Movement, 1890-1925*. Canadian Plains Research Centre: Regina. Canada.

Uma, Lele. 1984. *The Design of Rural Development : Lessons from Africa*. World Bank Research Publications. Washington.

Van, Blokland, P.J. *Introducing the Balance Sheet*. Extension Circular No. 651. Florida Cooperative Extension Service. Institute of Food and Agricultural Sciences. University of Florida. Gainesville. FL 32611.

Van, Blokland, P.J. *Introducing the Income Statement*. Extension Circular No. 645. Florida Cooperative Extension Service, Institute of Food and Agricultural Sciences. University of Florida. Gainesville. FL 32611.

Van, Blokland, P.J. *Introducing the Cash Flow*. Extension Circular No. 656. Florida Cooperative Extension Service. Institute of Food and Agricultural Sciences. University of Florida. Gainesville. FL 32611.

Van, Blokland, P.J. *Introducing the Statement of Change in Financial Position*. Extension Circular No.658. Florida Cooperative Extension Service. Institute of Food and Agricultural Sciences. University of Florida. Gainesville. FL 32611.

Van Blokland, P.J. *Linking the Financial Statements*. Extension Circular No. 657. Florida Cooperative Extension Service. Institute of Food and Agricultural Sciences. University of Florida. Gainesville. FL 32611.

Index